The Federal Role in Terrorism Insurance

Evaluating Alternatives in an Uncertain World

Lloyd Dixon, Robert J. Lempert, Tom LaTourrette, Robert T. Reville

RAND CENTER FOR TERRORISM RISK MANAGEMENT POLICY

The research described in this monograph was conducted within the RAND Center for Terrorism Risk Management Policy (CTRMP).

Library of Congress Cataloging-in-Publication Data

The federal role in terrorism insurance : evaluating alternatives in an uncertain world / Lloyd Dixon ... [et al.].
p. cm.
ISBN 978-0-8330-4235-4 (pbk. : alk. paper)
1. Terrorism insurance—Government policy—United States. 2. United States. Terrorism Risk Insurance Act of 2002. I. Dixon, Lloyd S.

HG8535.F43 2008
368.4'8—dc22

2007039017

The RAND Corporation is a nonprofit research organization providing objective analysis and effective solutions that address the challenges facing the public and private sectors around the world. RAND's publications do not necessarily reflect the opinions of its research clients and sponsors.

Published 2007 by the RAND Corporation
1776 Main Street, P.O. Box 2138, Santa Monica, CA 90407-2138
1200 South Hayes Street, Arlington, VA 22202-5050
4570 Fifth Avenue, Suite 600, Pittsburgh, PA 15213-2665
RAND URL: http://www.rand.org/
To order RAND documents or to obtain additional information, contact
Distribution Services: Telephone: (310) 451-7002;
Fax: (310) 451-6915; Email: order@rand.org

Preface

This monograph presents the findings of a RAND Center for Terrorism Risk Management Policy (CTRMP) study that sought to provide empirical evidence that could help address differences of opinion among stakeholders and within the federal government about many fundamental issues that are central to the current debate over extending the Terrorism Risk Insurance Act of 2002 (TRIA), as modified in 2005.

The monograph should interest those who want to better understand the potential consequences of allowing TRIA to expire at the end of 2007 on the take-up rate[1] for terrorism insurance and on the distribution across various segments of society of the losses that could result from a terrorist attack. It should also interest those who want to better understand the strengths and weaknesses of policy options for renewing TRIA, particularly on reforms intended to improve insurability against nuclear, biological, chemical, or radiological (NBCR) attacks. This monograph is also relevant to those interested in the application of robust decisionmaking (RDM) methods.

Expanding on the findings described in the documented briefing *Trade-Offs Among Alternative Government Interventions in the Market for Terrorism Insurance: Interim Results* (Dixon, Lempert, et al., 2007), this monograph draws on a body of RAND work related to TRIA and RDM, including the following:

- *Issues and Options for Government Intervention in the Market for Terrorism Insurance* (Dixon, Arlington, et al., 2004)
- *Distribution of Losses from Large Terrorist Attacks Under the Terrorism Risk Insurance Act* (Carroll et al., 2005)
- *Trends in Terrorism: Threats to the United States and the Future of the Terrorism Risk Insurance Act* (Chalk et al., 2005)
- *Shaping the Next One Hundred Years: New Methods for Quantitative, Long-Term Policy Analysis* (Lempert, Popper, and Bankes, 2003).

---

1 *Take-up rate* refers to the proportion of businesses that have insurance coverage for property losses resulting from terrorist attacks. As will be discussed further, workers' compensation (WC) policies always cover loss, regardless of cause.

The research reported here was supported by CTRMP as part of its larger research program focused on terrorism risk, insurance, and other economically focused issues related to terrorist threat.

The RAND Center for Terrorism Risk Management Policy

CTRMP provides research that is needed to inform public and private decision-makers on economic security in the face of the threat of terrorism. Terrorism risk insurance studies provide the backbone of data and analysis to inform appropriate choices with respect to government involvement in the market for terrorism insurance. Research on the economics of various liability decisions informs the policy decisions of the U.S. Congress and the opinions of state and federal judges. Studies of compensation help Congress to ensure that appropriate compensation is made to the victims of terrorist attacks. Research on security helps to protect critical infrastructure and to improve collective security in rational and cost-effective ways.

CTRMP is housed at the RAND Corporation, an international nonprofit research organization with a reputation for rigorous and objective analysis and the world's leading provider of research on terrorism. The center combines three organizations:

- RAND Institute for Civil Justice, which brings a 25-year history of empirical research on liability and compensation
- RAND Infrastructure, Safety, and Environment, which conducts research on homeland security and public safety
- Risk Management Solutions, the world's leading provider of models and services for catastrophe risk management.

For additional information about the Center for Terrorism Risk Management Policy, contact

Robert Reville
RAND Corporation
1776 Main Street
P.O. Box 2138
Santa Monica, CA 90407
Robert_Reville@rand.org
310-393-0411, x6786

Michael Wermuth
RAND Corporation
1200 South Hayes Street
Arlington, VA 22202
Michael_Wermuth@rand.org
703-413-1100, x5414

A profile of CTRMP, abstracts of its publications, and ordering information can be found at http://www.rand.org/multi/ctrmp/

Center for Terrorism Risk Management Policy Terrorism Insurance Project Advisory Board

Members with asterisks beside their names are also members of the CTRMP advisory board.

Jeffrey D. DeBoer (Cochair)*
President and chief executive officer
Real Estate Roundtable

Pierre L. Ozendo (Cochair)*
Member of the executive board
Head of Americas Property and Casualty
Swiss Re America Holding Corporation

Jack D. Armstrong*
Assistant vice president
Senior regulatory counsel
Liberty Mutual Insurance Company

Debra Ballen
Executive vice president
Public Policy Management
American Insurance Association

Richard A. Bayer
Executive vice president
Chief legal officer
The Macerich Company

Brian Boyden*
Executive vice president
State Farm Insurance

Timothy R. Campbell
Senior vice president
Government Relations
St. Paul Travelers

Andrew Coburn*
Vice president of catastrophe research
Director of terrorism research
Risk Management Solutions, Inc.

Bryon Ehrhart
President
Aon Re Services, Inc.

Kenneth Feinberg*
Managing partner
The Feinberg Group, LLP

Ann Spragens
Senior vice president and general counsel
Property Casualty Insurers Association of America

Paul L. Horgan
Partner
PricewaterhouseCoopers LLC

Paul Jardine
Chief executive
Catlin Underwriting Agencies Ltd.

Ken Jenkins*
Senior vice president
Chief underwriting officer
Specialty Markets
Munich Re America

Chris Lewis
Director of Alternative Risk Management Solutions
The Hartford Financial Services Group, Inc.

Peter Lowy*
Chief executive officer
Westfield Corporation, Inc.

Kathleen Nelson*
Immediate past chair
International Council of Shopping Centers

Art Raschbaum*
Executive vice president
Managing director
GMAC RE

Steve Sachs
Senior vice president
Managing director
National Real Estate Practice
Hilb Rogal and Hobbs

Jason Schupp
Assistant general counsel
Zurich North America

Kevin Scroggin*
General director
Corporate Risk Management and Insurance
General Motors

Hemant Shah*
President and chief executive officer
Risk Management Solutions, Inc.

Cosette Simon*
Senior vice president
Swiss Re Life & Health America Inc.

Richard Thomas*
Senior vice president
Chief underwriting officer
American International Group

Steven Wechsler*
President and chief executive officer
National Association of Real Estate Investment Trusts

Contents

Preface iii
Center for Terrorism Risk Management Policy Terrorism Insurance Project Advisory Board v
Figures ix
Tables xi
Summary xiii
Acknowledgments xxiii
Glossary xxv

CHAPTER ONE
Introduction 1
Federal Role in the Provision of Terrorism Insurance 1
Contributions of This Monograph 2
Organization of This Monograph 3

CHAPTER TWO
Analytic Methods 5
Approach to Uncertainty Analysis 6
Interventions, Uncertainties, Outcomes, and Relationships 8
Government Interventions 8
No Government Program 9
The Terrorism Risk Insurance Act 9
TRIA with NBCR-Attack Coverage 11
Outcome Measures 12
Key Uncertainties 15
Relationships in the Model 16
The Terrorist-Attack Model 17
The Take-Up Rate Model 20
The Postattack Government Compensation Model 23
The Insurance-Compensation and Loss-Distribution Model 24
Experimental Design 26

CHAPTER THREE
Consequences of Allowing TRIA to Expire ... 29
Take-Up Rates with and Without TRIA ... 30
Distribution of Losses with and Without TRIA ... 30
Conventional Attacks ... 30
NBCR Attacks ... 35
Trade-Offs Between Taxpayers and Insurers ... 37
Conventional Attacks ... 37
NBCR Attacks ... 39
A Probabilistic Look at the Trade-Offs in Outcomes for Large and Smaller Attacks ... 41
Cost to Future Policyholders ... 43
Summary ... 46

CHAPTER FOUR
Consequences of Requiring Insurers to Offer Terrorism Coverage for Conventional and NBCR Attacks ... 49
Take-Up Rate for a Policy That Covers Conventional and NBCR Attacks ... 50
Distribution of Losses for TRIA with NBCR-Attack Coverage ... 52
Conventional Attacks ... 52
NBCR Attacks ... 52
Cost to Future Policyholders ... 53
Trade-Offs Between Taxpayers and the Insurance Industry ... 59
Conventional Attacks ... 59
NBCR Attacks ... 60
Expected Taxpayer Cost ... 63
Extending TRIA to Cover NBCR-Attack Losses When the Cap Is Already Hard ... 65
Summary ... 66

CHAPTER FIVE
Conclusion ... 69
Key Findings ... 69
Implications of Findings for Recent Legislation ... 71
Moving Forward ... 72

APPENDIXES
A. RMS's Probabilistic Terrorism Model ... 73
B. Policyholder Take-Up of Terrorism Insurance ... 81
C. Loss-Distribution Model ... 99
D. Identifying Key Factors Driving Trade-Offs Between Interventions ... 107
E. Calculating Expected Losses with Multiple Probability Distributions ... 111

References ... 117

Figures

2.1. Schematic of Model 9
2.2. Loss Range Considered for Conventional Attacks........ 19
2.3. Loss Range Considered for NBCR Attacks 20
3.1. Distribution of Losses for Conventional Attacks with and Without TRIA for a Range of Assumptions, Including That About the Hardness of the Existing TRIA Cap........ 31
3.2. How Cost to Taxpayers and the Fraction of Losses Uncompensated Change for Three Example Scenarios When TRIA Is Allowed to Expire........ 34
3.3. Distribution of Losses for NBCR Attacks with and Without TRIA for a Range of Assumptions, Including That About the Hardness of the Existing TRIA Cap 36
3.4. Taxpayer Cost for Conventional Attacks with and Without TRIA........ 38
3.5. Share of Losses Covered by Insurers for Conventional Attacks with and Without TRIA........ 39
3.6. Taxpayer Cost for NBCR Attacks with and Without TRIA........ 40
3.7. Share of Losses Covered by Insurance for NBCR Attacks with and Without TRIA 40
3.8. Relationship Between Expected Annual Taxpayer Cost with and Without TRIA as the Probability of a Large Attack and Amount of Government Compensation Vary........ 42
3.9. Incidence of Losses Under TRIA, by Size of Attack, with $27.5 Billion Market Retention 44
3.10. Effect of Higher Recoupment on Taxpayer Cost and Cost to Future Policyholders for Conventional Attacks Under TRIA 45
4.1. Property Coverage Take-Up Rates for TRIA and TRIA with NBCR-Attack Coverage When the Hardness of the Existing TRIA Cap Is Uncertain........ 51
4.2. Distribution of Losses for Conventional Attacks Under TRIA with NCBR-Attack Coverage for a Range of Assumptions, Including That About the Hardness of the Existing Cap 54
4.3. Distribution of Losses for NBCR Attacks Under TRIA with NBCR-Attack Coverage for a Range of Assumptions, Including That About the Hardness of the Existing TRIA Cap 56

4.4. Distribution of Losses Under TRIA with NBCR-Attack Coverage, Hard Cap, and 7.5 Percent Deductible........ 58
4.5. Taxpayer Cost for Conventional Attacks Under TRIA and TRIA with NBCR-Attack Coverage, Hard Cap, and 7.5 Percent Deductible........ 60
4.6. Share of Losses Paid by Insurers for Conventional Attacks Under TRIA and TRIA with NBCR-Attack Coverage, Hard Cap, and 7.5 Percent Deductible........ 61
4.7. Taxpayer Costs for NBCR Attacks Under TRIA and TRIA with NBCR-Attack Coverage, Hard Cap, and 7.5 Percent Deductible........ 61
4.8. Share of Losses Paid by Insurers for NBCR Attacks Under TRIA and TRIA with NBCR-Attack Coverage, Hard Cap, and 7.5 Percent Deductible... 62
4.9. Expected Taxpayer Cost for TRIA Compared with That for TRIA with NBCR-Attack Coverage, Hard Cap, and 7.5 Percent Deductible........ 64
4.10. Take-Up Rates for TRIA and TRIA with NBCR-Attack Coverage When the Current Cap Is Hard........ 66
A.1. Exceedance Probability Distributions for Conventional and NBCR-Attack Losses........ 80
B.1. Approach Used to Calculate Change in Take-Up Rate........ 89
D.1. Coverage and Density Trade-Offs Offered by PRIM........ 108
E.1. Probability Distributions for Conventional and NBCR Attacks........ 112

Tables

2.1. Alternative Versions of TRIA with NBCR-Attack Coverage Analyzed........... 13
2.2. Outcome Measures Used to Compare Interventions................................ 14
2.3. Input Parameters Varied to Create the Ensemble of Plausible Futures............ 17
2.4. Attacks and Loss Ranges.. 19
2.5. Distribution of Losses with No Government Terrorism Insurance Program..... 25
2.6. Distribution of Losses Under TRIA-Based Intervention Structures 26
2.7. Numbers of Points in Experimental Designs....................................... 27
3.1. Summary of Outcomes with TRIA Relative to Outcomes Without TRIA...... 47
4.1. Comparison of Taxpayer Cost and Share of Losses Paid by Insurers Under TRIA and TRIA with NBCR-Attack Coverage.................................... 63
4.2. Summary of Outcomes for TRIA with NBCR-Attack Coverage, a Hard Cap, and a 7.5 Percent Deductible Relative to the 2007 Configuration of TRIA.. 67
A.1. Modes of Attack Modeled in the RMS Terrorism Risk Model.................... 75
A.2. RMS's Target Type Groups.. 75
B.1. Parameter Values Varied in Experimental Design to Determine Change in Take-Up Rate 90
B.2. Parameters That TRIA Expiration Affects.. 93
B.3. Parameters That Are Affected by Expanding TRIA to Require Offering Conventional and NBCR-Attack Coverage 94
C.1. Fraction of Loss Resulting from Fire 99
D.1. PRIM-Generated Clusters Explaining the 964/4,568 = 21 Percent of Cases in Which TRIA Imposes High Costs on Taxpayers....................... 109
D.2. PRIM-Generated Clusters Explaining the 2,348/4,568 = 51 Percent of Cases with a High Fraction of Unpaid Insured Loss If TRIA Expires.......... 109
D.3. PRIM-Generated Clusters Explaining the 1,177/4,568 = 26 Percent of Cases with Taxpayer Cost Under TRIA Higher Than That If TRIA Expires 110
E.1. Change in Expected Cost to Taxpayers If TRIA Expires........................ 113
E.2. Change in Expected Cost to Taxpayers for TRIA with NBCR-Attack Coverage, Low Deductible, and Hard Cap Versus TRIA 114
E.3. Change in Expected Cost to Taxpayers for Low Deductible, Hard Cap NBCR-Attack Offer Versus No Government Program........................... 115

Summary

Introduction

After the 9/11 attacks, most commercial insurers began excluding terrorism losses from their property-insurance policies, which previously had not identified the terrorism threat as a separate peril. Concerned that the unavailability of terrorism insurance would impede postattack economic recovery and hinder growth, Congress responded with the Terrorism Risk Insurance Act of 2002 (TRIA). TRIA requires commercial property-casualty insurers to offer insurance for losses suffered in terrorist attacks, and, in return, the federal government agrees to reimburse insurers for a proportion of claim payments that exceed a certain threshold amount. Congress amended TRIA in 2005, but TRIA will expire at the end of 2007 unless Congress takes further action.

After five years of support, the federal government's role in the market for terrorism insurance remains a subject of wide-ranging debate. Some envision TRIA as a temporary program needed only while insurers develop the tools and build the financial capacity to insure against terrorism risk. Others see a strong federal role in providing terrorism insurance as an ongoing necessity. Indeed, in the face of a growing awareness of the risk of nuclear, biological, chemical, and radiological (NBCR) attacks and routine exclusions of NBCR losses from coverage even when conventional attacks are insured, Congress is now considering not only whether to extend the program but also whether to expand it to improve the availability of NBCR coverage.

This monograph aims to contribute to this ongoing policy debate by addressing two directly relevant questions:

- What would be the implications of allowing TRIA to expire at the end of 2007?
- What would be the effects of modifying TRIA to improve the availability and affordability of insurance for NBCR attacks?

Analytic Approach

In answering these questions about the federal government's role in providing terrorism insurance, we used a simulation model to compare the outcomes of TRIA and potential alternative government interventions across a very wide range of plausible scenarios. We also used a robust decisionmaking (RDM) approach to handle the numerous, deep uncertainties that can confound any analysis of this topic. In the following sections, we discuss the program alternatives considered, the outcomes evaluated, the simulation model that relates the interventions to the outcomes, and the RDM approach used to exploit this model.

Program Alternatives Considered and Outcomes Evaluated

We considered three main alternative federal government interventions in the market for terrorism insurance:

1. TRIA as currently configured
2. no government program (the equivalent of allowing TRIA to expire)
3. TRIA modified to include NBCR coverage.

The first intervention replicates the current program as it exists in 2007, while the second intervention allows TRIA to expire. The TRIA-with-NBCR-coverage intervention modifies the current program by requiring insurers to offer policies that cover both conventional and NBCR attacks for a single price. In this alternative, policyholders must accept coverage for both conventional and NBCR coverage or decline terrorism coverage altogether. The monograph examines four variants of this intervention that differ in two key dimensions: the insurer deductible and the program cap.

The *TRIA deductible* refers to the maximum amount of insured losses for which insurers remain entirely responsible. Above the deductible, the federal government reimburses insurers for a proportion of their payments. This reimbursement comes from a mix of taxpayer funds and a federally mandated surcharge on future insurance policies. The higher the deductible, the greater an insurer's potential costs and, therefore, the more an insurer will charge for terrorism insurance. This monograph examines two levels of deductible for the TRIA-with-NBCR-coverage interventions: one identical to that of the current TRIA program, which is equal to 20 percent of an insurer's total written premiums on the insurance lines that the TRIA program covers, and one in which the deductible falls to 7.5 percent of insurer premiums.

The *TRIA program cap* refers to the provision in the TRIA legislation that limits total payments by insurers, future policyholders, and taxpayers for losses to no more than $100 billion. Because the legislation does not specify who is liable for losses beyond this amount, some insurers have expressed uncertainty about whether they would be directly or indirectly liable for any loss above the cap. Because many NBCR attacks could involve losses greater than $100 billion, an insurer's uncertainty about the "hardness" of the program cap increases its expected losses and, therefore, reduces its willingness to offer NBCR coverage. Thus, this monograph examines two types of caps for the TRIA-with-NBCR-coverage interventions: one that retains the current TRIA cap and one in which the cap becomes unambiguously binding (a "hard" cap). To harden the cap, we assume that the government would guarantee to pay all the insured losses from $100 billion to $650 billion.

For each of the TRIA, no-government-program, and TRIA-with-NBCR-coverage interventions, we evaluated performance using five outcome measures. Two measures—the fraction of losses that remain uncompensated after an attack and the cost to taxpayers—represent outcomes broadly reflecting impacts on society as a whole. Three measures—the fraction of insurance industry surplus used to compensate losses, the fraction of losses paid by the insurance industry, and the cost to future policyholders—represent outcomes reflecting the operation of the insurance marketplace and the role the insurance industry plays in bearing terrorism risk. Unlike previous studies, when calculating cost to taxpayers, we considered not only payments through the TRIA program but also payments made after an attack to provide compensation for uninsured losses or unpaid insured losses. We restricted our attention to property and workers' compensation (WC) losses. The TRIA program addresses other insured losses, such as losses on liability insurance policies, but they were beyond of the scope of this study.

We also examined the impact of the various government interventions on the take-up rate for terrorism insurance, which is an important intermediate variable that drives the five outcome measures considered. The *take-up rate* on property insurance policies refers to the proportion of property policies that have coverage for terrorism attacks and can differ for conventional attacks and NBCR attacks. The take-up rate for WC policies is 100 percent, because WC policies cover losses regardless of cause. When take-up rates are high, the insurance industry plays a larger role in compensating losses from terrorist attacks. When take-up rates are low, property owners' losses remain uncompensated unless the federal government pays them.

The Robust Decisionmaking Approach

Many of the most important underlying factors, or parameters, that determine the performance of the different government interventions, such as the frequency and magnitude of terrorist attacks, insurer beliefs about the hardness of the existing TRIA cap, and government assistance after an attack for businesses that fail to purchase terrorism insurance, are deeply uncertain. That is, there is no empirically based agreement on the value of these parameters or even the proper probability distribution to place over their plausible values. Policymakers and stakeholders implicitly make assumptions about these parameters that guide their decisions, but the assumptions can vary widely, contributing to vastly divergent views about the appropriate policies. Thus, this monograph considers the performance of the alternative government interventions over a very large number of plausible futures that capture a wide range of attacks, beliefs about the existing TRIA cap, levels of postattack government compensation, and other key factors. The monograph then uses RDM methods to identify patterns of outcomes generally observed across this broad range of futures and thus should help policymakers more confidently choose among the alternative government interventions despite the uncertainties involved.

The simulation model developed to evaluate outcomes over a wide range of futures includes a terrorist-attack model, a take-up rate model, a model of postattack government compensation, and an insurance-compensation and loss-distribution model.

- The terrorist-attack model predicts losses and probabilities of a large number of conventional and NBCR attacks of widely differing sizes.
- The take-up rate model predicts take-up for each of the TRIA-with-NBCR-coverage interventions based on the price of terrorism insurance, which is, in turn, determined by the cost to insurers of providing that insurance. We calibrate the take-up rate model to estimates in the literature for take-up rates with and without the current TRIA program.
- Given the uncertainty about what business assistance programs will be available after an attack, the postattack government compensation model considers levels of postattack government compensation that range from 0 percent to 75 percent of total uninsured and unpaid insured losses.[1]
- The insurance-compensation and loss-distribution model allocates losses caused by an attack across insurers, taxpayers, and businesses affected by the attack (in the form of uncompensated losses) and future insurance policyholders.

The hardness of the existing TRIA cap plays a key role in the both the take-up rate model and the insurance-compensation and loss-distribution model. For example, if the current cap is perceived to be very soft, hardening it will make a great deal of difference. Given the considerable uncertainty over the hardness of the cap, we consider scenarios in which the insurers are responsible for no insured losses and unpaid insured losses over the $100 billion TRIA program cap (a hard cap) up to scenarios in which insurers are responsible for 75 percent of such losses (a very soft cap).

---

[1] At the bottom of this range, government assistance is completely independent of the amount of uninsured loss and unpaid insured losses, which might be the case if government assistance were based only on the size of the attack and not the amount of uninsured losses. At the top of this range, the government will compensate most losses suffered by businesses without terrorism insurance. The higher the postattack compensation the government chooses to offer, the smaller the fraction of losses that are uncompensated but the higher the cost to taxpayers.

Key Findings

Using this analytic approach, we answered the two questions posed earlier.

Consequences of Allowing TRIA to Expire

TRIA has positive effects on the insurance market for *conventional* attacks relative to letting the program expire: The proportion of property-insurance policies with terrorism coverage is higher and the proportion of losses that remain uncompensated is lower for conventional attacks with TRIA than without TRIA.

TRIA's performance differs for larger and smaller conventional attacks. For conventional attacks with less than about $40 billion in total losses, TRIA increases the proportion of losses compensated by insurers relative to scenarios in which TRIA has expired and reduces taxpayer costs, once postattack government compensation is considered. For attacks with losses greater than about $40 billion, TRIA can reduce the role the insurance industry plays in compensating losses and can significantly increase the cost to taxpayers relative to scenarios without TRIA. For comparison, note that the attack on the World Trade Center caused roughly $23 billion in insured property and WC losses.

Even though TRIA saves taxpayers money only for conventional attacks causing less than $40 billion in damage, the expected annual taxpayer cost considering all types of attacks (conventional and NBCR) is lower with TRIA than without TRIA over a wide range of assumptions about the relative probabilities of large and small attacks and government compensation of uninsured and unpaid insured losses. This result holds because terrorism experts believe larger attacks to be far less likely than smaller ones. The higher taxpayer expense from government reimbursement in large and rare terrorist attacks is offset by the lower taxpayer cost in the likelier smaller terrorist attacks leading to net taxpayer savings. The costs are lower in the smaller attacks, because insurers, relieved of the risk from large attacks, offer lower prices, which increases take-up rates, lowering ex post government compensation and increasing the insurance share of compensation. The expected taxpayer costs remain lower under TRIA than without TRIA as long as the government compensates more than about 5 percent of uninsured and unpaid insured losses in the aftermath of an attack.

In contrast to the findings for conventional attacks, TRIA has done little to improve outcomes after NBCR attacks because of the continued low take-up rate for insurance coverage against NBCR attacks. More than 30 percent of the loss remains uncompensated in roughly 55 percent of the scenarios examined with or without TRIA in place. The primary benefit of TRIA for NBCR attacks is that the cap somewhat reduces the threat to the ongoing health of the insurance industry associated with large WC payouts. While this limited support of WC has value, many see the continued low take-up rate for property insurance against NBCR attacks as a significant gap in the nation's ability to manage terrorism risk.

Consequences of Expanding the Terrorism Risk Insurance Act of 2002 to Cover NBCR Attacks

To address the study's second question, we evaluated, as noted previously, four variants to TRIA that require insurers to offer bundled policies that cover terrorism losses due to both conventional and NBCR attacks—variants that differ in their deductibles and the hardness of their program caps. This analysis concluded that requiring terrorism policies to cover both conventional and NBCR attacks without changes in other program features such as the program cap or the insurer deductible may not improve outcomes much for NBCR attacks and may have significant adverse consequences for coverage of conventional attacks.

However, modifying the cap and deductible can improve outcomes for a program that requires insurers to offer both conventional and NBCR coverage. Specifically, hardening the cap and reducing the deductible from 20 to 7.5 percent generates outcomes comparable to those under TRIA in several key dimensions for scenarios associated with conventional attacks and significantly improves outcomes for scenarios associated with NBCR attacks. With such changes, the fraction of NBCR attack scenarios with uncompensated losses greater than 30 percent drops to only 11 percent compared to 56 percent under TRIA. This decline owes both to the higher take-up rates for NBCR coverage and government payment of all the insured loss between $100 billion and $650 billion.

Analogous to the finding for conventional attacks under the current version of TRIA, taxpayer cost is higher under TRIA with bundled NBCR coverage, a hard cap, and a 7.5 percent deductible than it is under TRIA for NBCR attacks that produce more than $40 billion in losses, but it is lower for many of the smaller NBCR attack scenarios examined. Because the probability of large attacks is perceived to be much lower than that of smaller ones, overall expected taxpayer cost is lower for a program that hardens the cap, lowers the deductible, and requires bundled NBCR coverage than it is for TRIA over a wide range of assumptions about the relative risk of large and smaller attacks and about the proportion of uninsured and unpaid insured losses compensated by the government. In this case, expected taxpayer cost will be lower given existing estimates of the relative probabilities of large and smaller terrorist attacks as long as the government compensates more than about 25 percent of uninsured and unpaid insured losses. Once again, government reimbursement of large losses in rare attacks lowers prices, which encourages NBCR take-up that reduces ex post government compensation in likelier smaller attacks.

Because of the uncertainty over the existing TRIA cap's hardness, our analysis suggests that both hardening the cap and lowering the deductible are critical to achieving positive outcomes when TRIA is expanded to require insurers to offer coverage for both NBCR and conventional attacks. If the existing cap is quite soft (that is, insurers may be liable for some fraction of losses above the $100 billion cap), lowering the deductible alone does not improve outcomes for NBCR attacks and can result in a deterioration of program performance for conventional attacks. If the cap is already fairly hard, hardening the cap would not make much difference, and lowering the deductible becomes key to avoiding adverse outcomes under TRIA. Hardening the cap while lowering the deductible is a robust strategy that effectively addresses the substantial uncertainty over how insurers perceive the hardness of the current cap.

Overarching Conclusions

Looking across the analysis of both questions addressed by this study, we found that, overall, both retaining TRIA and enhancing TRIA to cover NBCR attacks in a way that hardens the cap and lowers the deductible can achieve positive outcomes by transferring risk for the largest attacks to taxpayers. In return, the insurance industry can play a larger role in compensating losses for smaller attacks, and the resulting decline in uninsured losses means less government compensation after an attack. Because the probability of large attacks is thought to be far lower than the probability of smaller attacks, both TRIA and TRIA with NBCR coverage can achieve these benefits while reducing the expected taxpayer cost.

In choosing an extension to TRIA to better address NBCR attacks policymakers must be careful to choose an intervention that achieves the desired goals and avoids unintended consequences. For example, our analysis shows that simply extending TRIA to require a bundled offer of NBCR and conventional coverage without changing other program features, such as the cap or the deductible, can actually make the situation worse.

Implications for Recent Legislation

The U.S. House of Representatives has passed legislation that would extend and modify the TRIA program (H.R. 2761). The bill requires insurers to offer coverage for conventional and NBCR attacks, includes detailed language that attempts to harden the program cap, and lowers the deductible for NBCR attacks. While the interventions considered in this monograph differ in some important ways from this legislation, our analysis nonetheless provides some relevant insights.

The House bill includes several features identified in this monograph that will likely improve the performance of the TRIA program. First, the bill attempts to address the shortcomings of the TRIA program, identified here, for NBCR attacks. Second, the bill attempts to harden the TRIA cap, which our analysis suggests is important to successfully including NBCR coverage in the program. Finally, the bill lowers the deductible for NBCR attacks, consistent with the findings in our analysis.

Our analysis differs from the House bill in two important ways. First, the legislation attempts to harden the cap with detailed language and methods for prorating losses that exceed the cap, while the interventions considered here harden the cap by assuming the government guarantees to pay the insured loss more than $100 billion up to $650 billion. Critical to the House bill's impact on NBCR coverage will be insurers' perceptions about whether the bill's language is sufficiently strong to limit their actual liability for any insured loss that exceeds $100 billion and how these perceptions evolve over time.

Second, the House bill links offers for NBCR and conventional terrorism coverage differently from how the options considered in this study do so. As in our analysis, the House bill requires insurers to offer coverage for conventional and NBCR attacks that does not differ in terms, amounts, or other conditions for coverage for events other than terrorism. We require policyholders to either accept or reject this bundled coverage. Under the House bill, in contrast, if the policyholder rejects the initial offer of coverage, the insurer may offer coverage options that differ in terms, amounts, or other conditions from the underlying policy. In particular, an insurer may offer coverage for only conventional attacks and not NBCR attacks.[2] The question remains whether allowing policyholders to separately purchase conventional and NBCR coverage will result in a sufficiently high take-up rate for NBCR coverage to generate outcomes similar to those found in this analysis. Existing research suggests that the demand for NBCR coverage is low; thus, allowing this coverage to be offered separately may not result in substantial take-up. Further research on the effect of offering unbundled versus bundled conventional and NBCR coverage is clearly warranted. However, given the potential importance of this issue and the shortage of solid evidence on which to base any judgments, Congress should plan to review the effects of new legislation on NBCR take-up and revise its approach in the next few years as appropriate, even if it chooses to reauthorize the overall TRIA program for a longer period.

Moving Forward

This monograph does not address some issues relevant to a full assessment of government intervention in the market for terrorism insurance. For instance, we do not assess (1) the impact of changes in insurance price and take-up rate caused by TRIA and enhancements to it on economic activity preattack or the speed of economic recovery and resiliency of the economy after an attack, (2) how price changes might affect incentives for businesses to adopt measures to mitigate terrorism risk, (3) how any change in the insurance industry's existing willingness to bear terrorism risk might affect take-up rates over time, or (4) how government programs affect the flow of new capital into insurance markets or the development of instruments or strategies to spread insurance

[2] That we modeled a different approach than the House bill should signal no policy preference. We settled on the options analyzed here before the House bill was introduced and chose to bundle NBCR and conventional coverage because (1) it seemed like the simplest way to extend TRIA to better address NBCR attacks, and (2) it is analytically more straightforward to analyze bundled coverage than to analyze unbundled coverage.

risk. Including such issues would increase the comprehensiveness of our analysis but would not affect the basic trade-offs identified here.

The models and methods used in this study apply to a wider range of questions than those considered here. For instance, our simulations could be adapted to examine a broader range of modifications to the TRIA program, such as different insurer copayments or different program caps. These tools could also examine a broader range of government interventions in terrorism-insurance markets, including requiring policyholders to purchase terrorism coverage or pooling arrangements in which a surcharge on insurance policies funds a pool that is then used to pay claims following a terrorist attack.

The threat of terrorism does not appear to be a transitory phenomenon confronting the United States, and the role that insurance can play in mitigating this threat warrants ongoing analysis.

Acknowledgments

We would like to thank the project advisory board for extremely helpful assistance and feedback during the course of the project. Advisory board members contributed their expertise on the issues relevant to terrorism risk and insurance through interviews, written comments, and attendance at multiple advisory board meetings.

Three reviewers provided insightful comments as part of the formal RAND peer-review process: Eric Helland, jointly at RAND and Claremont McKenna College; Richard Hillestad at RAND; and George Zanjani at the Federal Reserve Bank of New York. Howard Kunreuther and Erwann Michel-Kerjan of the Wharton Risk Management and Decision Processes Center at the University of Pennsylvania also provided detailed and helpful comments on interim results. We thank them all for their time and the speed with which they turned around their comments.

Access to Risk Management Solutions's (RMS's) Probabilistic Terrorism Model was critical to the success of the project, and we thank RMS for making the model and its expertise available. We also thank Evolving Logic for the use of its Computer Assisted Reasoning System (CARs™) software, on which we relied heavily in this analysis.

At RAND, Scot Hickey skillfully ran the RMS model, Benjamin Bryant investigated the factors leading to the vulnerabilities of the alternative government interventions examined, Lisa Bernard did an excellent job editing and formatting the document under tight deadlines, Joye Hunter efficiently coordinated the review process, and Stacie McKee did a superb job moving the document quickly though the publication process. Rebecca Collins ably headed RAND's quality assurance process for this monograph, and Michael Wermuth provided useful suggestions throughout the project. Their efforts are greatly appreciated.

Glossary

CARs™	Computer Assisted Reasoning System
conventional attack	a terrorist attack that does not involve or trigger nuclear, biological, chemical, or radiological reaction, release, or contamination (see also *NBCR attack* below)
CTRMP	Center for Terrorism Risk Management Policy
DEP	direct earned premium
future	A set of values for the uncertain model input parameters; the combination of a future with a particular government intervention yields a *scenario.*
insurer copayment	the proportion of insured losses above the insurer deductible and below the TRIA cap that the insurer must pay
insurer deductible	claim payments that an insurer must make after an attack before any reimbursement from the federal government is made
insurer surplus	the difference between an insurer's assets and its liabilities; the financial cushion that protects policyholders in the case of unexpectedly large claim costs
LHC	Latin hypercube
NBCR	nuclear, biological, chemical, or radiological
NBCR attack	a terrorist attack that involves or triggers nuclear, biological, chemical, or radiological reaction, release, or contamination
PRIM	patient rule induction method

RDM	robust decisionmaking
reinsurer	a company that provides insurance to primary insurers
reinsurer capacity for conventional attacks (or for nuclear, biological, chemical, or radiological attacks)	the total amount of coverage that reinsurers are willing to write for conventional attacks (or for nuclear, biological, chemical, or radiological attacks)
RMS	Risk Management Solutions, Inc.
scenario	a combination of a government intervention and a future (see also *future*)
take-up rate	the proportion of policyholders that have coverage
TRIA	Terrorism Risk Insurance Act of 2002
TRIA deductible	the maximum amount of insured losses for which insurers remain entirely responsible; above this amount, the federal government reimburses insurers for a proportion of their payments.
TRIA industry market retention	the amount of insured loss that the insurance marketplace must pay through insurance claim payments and future commercial policyholder surcharges before taxpayers begin to contribute to paying the insured loss
TRIA program cap	the amount of total annual insured losses in insurance lines that TRIA covers above which neither an insurer nor the government is liable for paying claims
WC	workers' compensation
unpaid insured losses	losses that are insured but that insurers do not pay either because of the TRIA program cap or because the insurance industry surplus is exhausted
XLRM factors	uncertainties (X), policy levers (L), relationships (R), and measures (M)

CHAPTER ONE

Introduction

The proper federal role in the provision of insurance against terrorist attacks is the subject of a wide-ranging debate. After the 9/11 attacks, most commercial insurers began excluding losses caused by terrorist attacks from their property-insurance policies, which previously had not identified terrorism as a separate peril. Concerned that this unavailability would impede economic recovery after the attacks and hinder growth, Congress responded with the Terrorism Risk Insurance Act of 2002 (TRIA). The act requires commercial property-casualty insurers to offer insurance for losses suffered in terrorist attacks, and, in return, taxpayers and holders of future commercial insurance policies will reimburse insurers for a proportion of claim payments exceeding a certain threshold amount.[1] Some envision TRIA as a temporary program needed only while insurers develop the tools and build the financial capacity to insure against terrorism risk. Others see a strong federal role in the provision of terrorism insurance as an ongoing necessity.

The current TRIA program expires at the end of 2007, and Congress is currently debating whether to extend the program and, if so, whether it should be modified. This monograph aims to contribute to that debate by addressing two questions: First, what would be the implications of allowing TRIA to expire in the aftermath of any future terrorist attack? Second, what would be the effects of expanding TRIA to improve the availability and affordability of insurance for nuclear, biological, chemical, and radiological (NBCR) attacks?

Federal Role in the Provision of Terrorism Insurance

The immediate aftermath of 9/11 saw little debate over at least a temporary need for a program such as TRIA. Insurers dropped terrorism from their coverage, and, when such coverage was available, prices tended to be very high (GAO, 2002; Saxton, 2002). The resulting lack of coverage led to fears that the national economy would suffer. Anecdotal evidence suggested that the lack of coverage in the face of new requirements from

[1] TRIA does not apply to homeowners' and other types of personal (as opposed to commercial) insurance.

commercial lenders and investors to secure terrorism insurance discouraged building and leasing, particularly in large metropolitan areas (GAO, 2002; Saxton, 2002).

In the roughly five years that TRIA has existed, the insurance industry appears to have improved its ability to supply terrorism coverage. Market capacity and policyholder take-up of such insurance have increased, and some argue that continuation of the program could hinder further development of the market (PWG, 2006; Torregrosa, 2005). However, there are also reasons to believe that a private market for terrorism insurance cannot function properly without a strong federal role. Reasons typically cited (see, e.g., Jaffee and Russell, 2005; Brown et al., 2004; Dixon, Arlington, et al., 2004; Saxton, 2002) include the belief that the risk is too uncertain to allow meaningful pricing and that losses could exceed industry net worth. A strong federal role may also be appropriate because U.S. foreign policy may, to some extent, determine the risk, because relevant information for managing it likely remains classified and because it may be more equitable to spread the risk across the nation as a whole than to let it fall more heavily on particular property owners.

Researchers, policymakers, and industry representatives continue to debate ways in which TRIA has helped and hindered the post-9/11 development of markets for terrorism insurance and whether the private sector can ever provide such insurance without an enduring federal role (see Jaffee and Russell, 2005; Brown et al., 2004; PWG, 2006; Cummins, 2006; Barker, 2003; Hubbard, Deal, and Hess, 2005; Torregrosa, 2005, 2007). In addition, TRIA does not require coverage for NBCR attacks, and little reinsurance for such coverage is available (U.S. Government Accountability Office and U.S. House of Representatives, 2006; PWG, 2006).[2] Some believe TRIA ought to be expanded to include such perils.

It is within the context of this debate over the proper federal role in the provision of terrorism insurance that policymakers are considering the renewal of TRIA. Some specific proposals are emerging, including a bill from the House of Representatives (U.S. House of Representatives, 2007) that would renew TRIA for 15 years and expand the program to require insurers to offer coverage for NBCR attacks.

Contributions of This Monograph

Any analysis of the federal role in the provision of terrorism insurance must contend with a wide range of difficult-to-estimate uncertainties, including what types of terrorist attacks might occur, how the industry might respond to alterative government interventions, and what decisions the government might make after an attack regarding compensation of the uninsured and unpaid insured loss. These questions all repre-

---

[2] Estimates of reinsurer capacity for conventional attacks run from \$6 billion to \$8 billion and from \$0.9 billion to \$1.6 billion for NBCR attacks (PWG, 2006, p. 26).

sent what we have called "deep uncertainty," in which not only the outcomes, but also the probability of different outcomes, are difficult to estimate with any accuracy (see Lempert, Popper, and Bankes, 2003).

This monograph takes a more systematic approach to addressing this deep uncertainty than have previous analyses of the TRIA program. Using robust decisionmaking (RDM) methods (Lempert, Popper, Bankes 2003), our analysis compares TRIA and potential alternative government interventions across a very wide range of plausible scenarios. We developed a computer simulation model that estimates the distribution of losses among property owners, insurers, taxpayers, and others, given an intervention and any of a wide range of assumptions about the type of terrorist attack, preattack behavior of the industry, and postattack compensation decisions of the government. The government interventions considered include letting TRIA expire and expanding TRIA to require insurers to offer policies that cover both conventional and NBCR attacks. We identify and characterize the key trade-offs among the alternative government interventions.

This analysis builds on two earlier studies that examined the distribution of losses expected under TRIA and potential alternatives (Carroll et al., 2005; Kunreuther, 2005). Our study offers two primary advancements in addition to its more systematic consideration of uncertainty. First, we developed a model of policyholder take-up of commercial property insurance for terrorism. The model links changes in the government program to the cost, and subsequently the price, of terrorism insurance. Changes in the price induce changes in policyholder take-up of terrorism coverage, which is critical to understanding changes in the role that the insurance industry plays in spreading terrorism risk and the proportion of the loss after an attack that potentially goes uncompensated. Second, the analysis considered potential government decisions about compensation of uninsured or unpaid insured losses after an attack. Many analyses ignore such postattack government compensation because it is difficult to predict, but it nonetheless can prove crucial in the comparison of alternative preattack government programs.

Organization of This Monograph

Chapter Two summarizes the analytical approach employed in this monograph, including a description of the alternative government interventions examined, the outcome measures used to compare them, the modeled losses from terrorist attacks, the policyholder take-up model, and the RDM approach. The appendixes provide further details of some aspects of the analysis. Chapter Three examines the trade-offs between TRIA and allowing TRIA to expire. Chapter Four compares the current TRIA program to modified versions that require insurers to offer policies that provide coverage for both conventional and NBCR attacks. Chapter Five summarizes our results, highlights the trade-offs among alternative interventions, and suggests how this analysis might apply to the proposed renewal of TRIA currently before the House of Representatives.

CHAPTER TWO

Analytic Methods

This chapter summarizes the approach used in this analysis. Our approach involves constructing simulation models to estimate the effects of different government interventions in the market for terrorism insurance and then using an RDM approach to examine the trade-offs among alternative options in terms of specific outcome measures.

The simulation model projects how alternative government interventions in the terrorism-insurance system perform according to several outcome measures over a wide range of plausible futures. The outcome measures we use are related to the distribution of losses among various parties. The simulation projects the losses from a range of terrorist attacks, the effects of different government intervention options on terrorism-insurance take-up among commercial policyholders, and those options' effects on the distribution of losses among various stakeholders to calculate the value for each outcome measure. Our model includes the possibility for postattack government compensation of both uninsured and unpaid insured losses under all intervention options considered.

RDM methods, explained in more detail later in this chapter, can systematically compare the alternative interventions even when many important input parameters to the model are deeply uncertain. The approach entails running the model over a wide range of parameter values and identifying combinations of these inputs (e.g., size of attack, type of intervention) that lead to scenarios with significant, policy-relevant differences among outcome measures (e.g., distinguish among scenarios that cost or save taxpayer money). RDM helps identify robust interventions—that is, those interventions that result in the most desirable outcomes over the widest range of potential futures.

This chapter provides an overview of

- our approach to uncertainty in the models and underlying parameters
- the government interventions considered in the market for terrorism insurance
- the outcome measures used to evaluate the interventions' performance
- the key uncertainties addressed in the analysis

- the models of the terrorism-insurance market and for the distribution of the losses resulting from a terrorism attack across the various stakeholder groups
- the methods used to construct the scenarios used in the analysis.

The appendixes provide more detail on these topics.

Approach to Uncertainty Analysis

The traditional policy analysis approach for assessing alternative federal government interventions in terrorism-insurance markets would rest on a probabilistic assessment of the likelihood of various types of terrorist attacks and other key uncertainties. The analysis would use some type of system model that describes outcomes of interest contingent on the choice of government intervention. The analysis would then recommend the government action with the optimal expected utility, contingent on these distributions. In sophisticated applications, sensitivity analysis (Saltelli, Chan, and Scott, 2000) can then suggest how different assumptions about parameter values or probability distributions might affect this ranking of policies.

The traditional optimum-expected-utility approach to risk management has proved extraordinarily useful in a wide range of decision challenges. (See Morgan and Henrion, 1990, for an excellent review.) However, it has significant shortcomings when applied to terrorism insurance, because the problem affects a large number of diverse interests and presents uncertainty so large that it is not possible to confidently define a system model or prior probability distributions of the inputs. We use the term *deep uncertainty* to describe such conditions, defined as a situation in which decisionmakers do not know or cannot agree on the system model that relates action to consequences, prior probabilities of the inputs to the system model(s), or value functions that rank the desirability of the consequences (Lempert, Popper, and Bankes, 2003). If applied under such conditions of deep uncertainty, traditional optimum-expected-utility methods can encourage analysts and decisionmakers to be sufficiently overconfident in their estimates of uncertainty to make predictions more tractable; can make agreement on actions more difficult, as parties gravitate toward the differing expert pronouncements of probability distributions most compatible with their own individual values, policy priorities, or decision contexts; and can lead to strategies vulnerable to surprises that might have been countered had the available information been used differently (Groves and Lempert, 2007).

In recent years, several analytic approaches have been developed to address some of these problems of decisionmaking under such conditions of deep uncertainty (Ben-Haim, 2001; Rosenhead, 2001; Levin and Williams, 2003). RDM is one such approach (Lempert, Groves, et al., 2006; Lempert and Popper, 2005; Lempert, Popper, and Bankes, 2003) that attempts to characterize uncertainty with multiple representations of the future rather than a single set of probability distributions[1] and by using robustness, rather than optimality, as a decision criterion (Lempert and Collins, forthcoming). RDM considers robust strategies as ones that perform relatively well, compared to alternative strategies, across a wide range of plausible future states of the world.[2] RDM is consistent with traditional optimum-expected-utility analysis in the sense that, in the limit of the multiple futures reducing to a single probability distribution, the optimum and robust solutions become the same. In contrast to traditional sensitivity analysis, which often suggests how the ranking of strategies may change with differing assumptions, RDM seeks to identify strategies whose satisfactory performance, compared to the other strategies, is relatively insensitive to all or most of the most significant uncertainties.

This monograph does not provide a full RDM analysis, since it does not offer any overall assessment of the robustness of alternative government interventions. In part, our ability to do so is limited because we do not consider how the government interventions might affect total societal losses due to terrorism, for instance, by influencing firms' decisions on location or security measures or by influencing terrorists' attack plans. Rather, the analysis focuses on how the loss from any given attack is distributed among different segments of society (such as property owners, the insurance industry, and taxpayers) and the role that private insurers play in underwriting terrorism risk.

This monograph focuses on a key step in an RDM analysis: identifying the strengths and weaknesses of alternative strategies and characterizing the key trade-offs among them.[3] We thus emphasize exploratory modeling methods (Bankes, 1993), which use databases created from multiple runs of computer simulation models to systematically explore the implications of a wide variety of assumptions and hypotheses. In particular, we generate an experimental design over multiple uncertain input parameters to a computer simulation model to create an ensemble of plausible futures against which we "stress test" alternative government interventions. We then use summary visualizations and statistical analyses to illuminate the key trade-offs among the alternative interventions and identify the key factors driving these trade-offs.

For instance, in Chapter Three, we compare the cost to taxpayers with and without TRIA for each scenario in the ensemble of plausible futures.[4] We apply statistical cluster-finding algorithms that suggest that the size of the attack represents the key

---

1 Such multiple views of the future are used in a variety of policy areas—for example, those that consider monetary policies robust over competing economic models (Levin and Williams, 2003).

2 See Lempert and Collins (forthcoming) for a formal definition of *robust strategies.*

3 These steps represent a quantitative implementation of the qualitative assumption–based planning method (Dewar, 2002).

4 We define *scenario* as the combination of a specific government intervention and a particular realization of the uncertain model-input parameters. A particular realization of the uncertain model-input parameters alone is referred to as a *future.*

factor in determining whether the cost to taxpayers with TRIA exceeds the cost if TRIA expires. This analysis suggests that the former cost will exceed the latter only for attacks causing more than roughly $40 billion in damage.

This approach also allows us to appropriately include uncertain probabilistic information in the analysis. We use such information in two ways. First, as described below, we use ex ante estimates of the probability distribution of losses from terrorist attacks in the model of take-up rates for terrorism insurance. Second, we use probabilistic information as reference points when we compare the postattack (ex post) vulnerabilities of alternative interventions. For instance, we calculate the expected value of the cost to taxpayers under TRIA for a wide range of exceedance curves describing the probability of a large, conventional terrorist attack. We note that the probability of such attacks may need to be roughly an order of magnitude larger than that estimated by the Risk Management Solutions (RMS) model for the expected cost to taxpayers under TRIA to exceed that if TRIA expires.

Interventions, Uncertainties, Outcomes, and Relationships

In describing the computer simulation used for an RDM analysis, it is often useful to organize the contributing factors into four categories: *uncertainties* outside of decisionmakers' control, *policy interventions* under consideration by decisionmakers, *outcome measures* that decisionmakers and other interested parties will use to rank the desirability of various scenarios, and the *relationships* that govern how the policies and uncertainties affect those attributes of the system related to the measures.[5] Figure 2.1 displays the simulation model used in this monograph in these categories. The government interventions, outcome measures, and uncertainties are shown in octagons, and the modules representing the relationships are shown in shaded boxes. This chapter now describes each in turn.

Government Interventions

Our analysis examined several alternative government interventions in markets for terrorism insurance. These alternatives do not represent specific proposals of any stakeholder to the TRIA debate, nor do we intend them as policy options that might be adopted. Rather, these interventions represent a diversity of approaches the federal government might consider and were chosen to illuminate the trade-offs implicit in the choice of different combinations of elements that might go into any actual

---

[5] Lempert, Popper, and Bankes (2003) describe these XLRM factors—uncertainties (X), policy levers (L), relationships (R), and measures (M)—in detail.

Figure 2.1
Schematic of Model

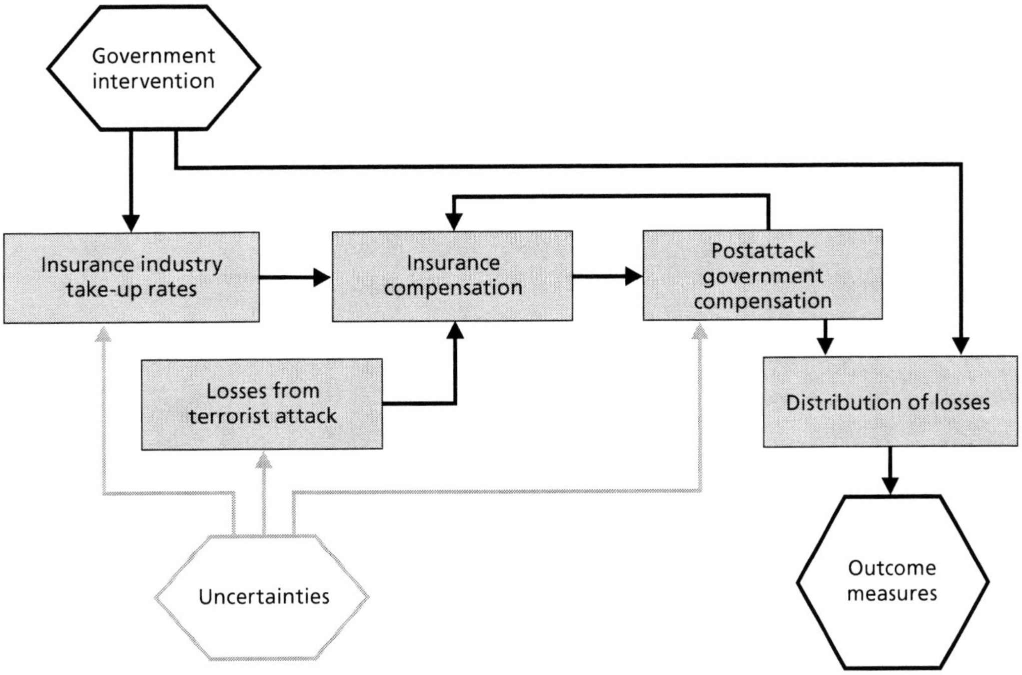

RAND *MG679-2.1*

government program. This monograph's comparison of these alternative interventions aims to guide the design of any actual government intervention.

We consider the following alternative government interventions:

1. no government program (i.e., letting TRIA expire)
2. TRIA as configured in 2007
3. TRIA expanded to address NBCR attacks, including four combinations of the insurer deductible and program cap.

No Government Program

In this alternative, no federal government intervention exists in the market for terrorism insurance. Workers' compensation (WC) coverage is still subject to state regulations requiring coverage for all losses, including those from terrorism.

The Terrorism Risk Insurance Act

In this alternative, TRIA as it exists in 2007 is extended indefinitely.

TRIA requires property-casualty insurers to make terrorism coverage available to their commercial policyholders on the same terms as their coverage for losses from other perils.[6] In return, the federal government agrees to reimburse insurers for a proportion of payments to commercial policyholders for losses from terrorism attacks that exceed certain thresholds. Insurers set insurance premiums subject to state regulation that may be applicable, and purchase by policyholders remains voluntary.[7] TRIA applies to most commercial property-casualty lines, including WC insurance, but not to health, life, or personal insurance lines such as home or auto. The U.S. Department of the Treasury administers the program.

Several TRIA features—in particular the insurer deductible, program cap, and recoupment—play important roles in our analysis.

Under TRIA, an insurance company is responsible for all insured losses up to that company's TRIA deductible. The government then covers 85 percent of the insurer's payments until the combined payments for all insurers and the public equal $100 billion. The insurer's deductible is a fraction of its direct earned premium (DEP) in the insurance lines subject to TRIA in the previous year, which is 20 percent in 2007.[8] The U.S. Department of the Treasury estimated that the premiums in TRIA lines totaled $182 billion in 2006, meaning that the industrywide deductible could amount to as much as $36 billion in 2007 (assuming that the loss affects all insurers in proportion to premiums earned).

TRIA states that neither the government nor an insurer that has met its insurer deductible shall be liable for the payment of aggregate insured losses that exceed $100 billion (we refer to the $100 billion threshold as the *program cap*). However, the act does not specify whether or how the insured loss over $100 billion will be paid, stating instead that "Congress shall determine the procedures for and the source of any payments for such excess insured losses" (P.L. 107-297, §103[e][3]).[9] Some of the insurers we interviewed during the course of this project reported uncertainty about whether they would be directly or indirectly liable for the loss above this threshold.[10] Conse-

6 Congress passed TRIA in November 2002 (P.L. 107-297) and extended it with the Terrorism Risk Insurance Extension Act in 2005 (P.L. 109-144). The law is set to sunset on December 31, 2007. Several authors have published good descriptions of TRIA's structure (e.g., Carroll et al., 2005; Chalk et al., 2005; Kunreuther and Michel-Kerjan, 2006), and readers are referred to these publications for more details.

7 Insurers must initially offer terrorism coverage up to the same limits as in the underlying insurance policy but are free to set the price. If the policyholder declines the offer, the insurer can come back with a different offer, such as one with a lower coverage limit and a lower price.

8 In general, a *direct earned premium* is the portion of gross premium collected from a policyholder before the deduction for reinsurance premiums for which all or part of the insurance policy term has expired. (Insurance premiums are payable in advance, but the insurance company earns them only as the policy period expires and in proportion to the expired period.) Under treasury department regulations issued to implement TRIA, the direct earned premium is the premium information that insurers report in column 2 of the "NAIC Exhibit of Premiums and Losses of the Annual Statement" (commonly known as Statutory Page 14). See 31 CFR 50.5(d)(1).

9 For more information on TRIA, see U.S. Department of the Treasury (2006).

10 CBO made a similar finding (Torregrosa, 2007, p. 14). Insurers could be directly liable if state courts or the U.S. Congress ordered them to pay all insured losses after an attack. They could be indirectly liable if, for exam-

quently, when examining the existing program's performance, we allow the fraction of the insured loss over the cap paid by insurers to vary from 0 to 75 percent. When the fraction is low, the cap is nearly hard. When it is at the upper end of the range, the cap is quite soft. To better understand outcomes when the cap is hard, we also separately examined scenarios in which insurers are not liable for insured losses after the $100 billion cap is reached.

The government charges no premium for the coverage that it provides under TRIA. The act does require the government, after any attack, to recoup some or all of its outlays through subsequent surcharges on commercial property-casualty premiums. In 2007, the government must recoup payments until the total amount paid by insurers plus the policyholder surcharge equals $27.5 billion. The act gives the government discretion to decide after an attack whether to recoup a larger amount of government payments to insurers.

TRIA with NBCR-Attack Coverage

TRIA requires that terrorism coverage be offered under the same terms, amounts, and limitations as those applicable to losses arising from events other than terrorism. Because commercial property policies nearly always exclude losses resulting from NBCR attacks, terrorism coverage also excludes property losses from NBCR attacks in most cases.[11] As a result, TRIA does little to provide insurance for property loss from NBCR attacks, although state regulations do require that WC policies cover loss regardless of how it occurs.

This monograph examines four variants of the TRIA-with-NBCR-coverage intervention that aim to improve the program's ability to address NBCR attacks. The interventions leave the program's basic structure unchanged but expand it by removing NBCR exclusions for terrorism coverage. Our analysis considers terrorism policies that bundle conventional and NBCR coverage. In these interventions, the government requires that the mandatory offer of terrorism coverage include a rider lifting the NBCR exclusion for terrorism so that all policies cover both conventional and NBCR attacks. Policyholders cannot select only conventional or only NBCR coverage; they must either accept coverage for both types of attacks or decline terrorism coverage altogether.[12]

---

ple, Congress decided to impose taxes on insurers to fund unpaid claims.

[11] Many states mandate that losses from fire be covered even if the fire was caused by a peril that is not covered (so-called "fire-following" coverage). Thus, property losses caused by NBCR attacks will be covered in some circumstances. For example, if a radiological bomb caused a fire when it detonated, the property damage due to fire would be covered, but damage would not be covered that was due to radiological contamination.

[12] As will be discussed in Chapter Five, the TRIA bill recently passed by the U.S. House of Representatives requires insurers to offer coverage for conventional and NBCR attacks but, in effect, allows policyholders to choose coverage for one risk but not the other. We settled on the options analyzed here before the House bill was introduced and chose to bundle NBCR and conventional coverage (1) because it seemed like the simplest way to

We explore two features critical to the performance of such TRIA extensions that address NBCR: the hardness of the program cap and the size of the insurer deductible. What insurers believe about the cap will likely strongly influence the impact of expanding TRIA to require policies to cover both conventional and NBCR attacks. The increase in premium and the decline in take-up rate will be greater if insurers believe that the current TRIA cap is soft than it will be if they believe it is already hard. The TRIA deductible governs the industry share of the loss below the program cap. Including NBCR in the program will increase insurers' expected loss. Lowering the deductible will lower insurers' expected loss. The government might choose to lower the deductible if adding NBCR to TRIA.

This monograph thus considers four variants of TRIA with NBCR coverage, as shown in Table 2.1, that explore combinations of (1) the existing cap (with its ambiguity over insurer responsibility for insured losses that exceed $100 billion) and a hard cap (with no such ambiguity) and (2) of the current (20 percent) and a lower (7.5 percent) TRIA deductible. To harden the TRIA cap, we chose a straightforward, though perhaps politically unrealistic, approach: government guarantee of payment for all of the insured loss above the TRIA cap up to $650 billion. As shown in the third column of Table 2.1, when the language concerning the cap in the TRIA legislation remains unchanged, we model scenarios in which insurers both expect to pay and end up paying anywhere from 0 to 75 percent of insured losses that exceed $100 billion. We chose to reduce the deductible to 7.5 percent in the low-deductible variants because a 7.5 percent deductible was included in the House's ultimately unsuccessful TRIA reform bill in 2005.

Outcome Measures

Given a government intervention, the simulation model used in this monograph allocates the financial loss from an attack across five categories: private commercial property and WC insurers (insurance industry), which pay losses from their surplus; taxpayers, who are assumed ultimately responsible for any cost borne by the federal government; all private commercial property and WC insurance policyholders nationwide (future policyholders), who may be assessed a surcharge to compensate past losses; owners of uninsured commercial property and uninsured businesses in those properties (uninsured loss); and insured commercial property owners, businesses, and employees that do not receive insurance payments, either because of insurance company failures or

extend TRIA to better address NBCR attack and (2) because it is analytically more straightforward to analyze a bundled policy than it is to analyze an unbundled policy.

Table 2.1
Alternative Versions of TRIA with NBCR-Attack Coverage Analyzed

	Percentage					
	Insurer Share			Government Share		
Alternative	Deductible (percent of prior-year DEP)	Copayment up to $100 Billion TRIA Cap	Payments from $100 Billion to $650 Billion	Copayment up to $100 Billion TRIA Cap	Payments from $100 Billion to $650 Billion[a]	Compensation Above $650 Billion
Existing cap, high deductible	20	15	0–75	85	0–75	0–75
Existing cap, low deductible	7.5	15	0–75	85	0–75	0–75
Hard cap, high deductible	20	15	0	85	100	0–75
Hard cap, low deductible	7.5	15	0	85	100	0–75

[a] After insurer payments have been made.

because the total insured loss exceeds a program threshold (unpaid insured loss).[13] The analysis uses the distribution of losses among these categories to compare the performance of the government interventions.

Ideally, our analysis would allow us to assess the alternative government interventions according to their effects on social welfare, broadly defined, before and after any terrorist attack. Before an attack, government interventions ought to be assessed according to the extent to which they encourage firms to reduce their vulnerability to terrorist attacks, the extent to which terrorists are deterred from conducting attacks, the development of new instruments and capacity for bearing terrorism risk, and the effect on the growth of jobs and income. After an attack, government interventions ought to be assessed according to the extent to which they facilitate recovery, maintain a well-functioning insurance marketplace, and minimize the costs of response, as well as according to the resulting distribution of losses across various segments of society.

Rather than address this full range of societal impacts, however, this monograph focuses on how a particular government intervention affects the allocation of losses from any given terrorist attack among the five categories mentioned above. We combine these model outputs into five outcome measures for the alternative government interventions. Each serves as a direct measure of the performance of the government intervention and as a proxy for a broader social welfare concern (Table 2.2).

The first two measures broadly represent social outcome measures, while the last three represent outcome measures more focused on the operation of the insurance marketplace. Each is discussed in more detail below.

Table 2.2
Outcome Measures Used to Compare Interventions

Outcome Measure	Importance
Fraction of loss that remains uncompensated after an attack	Focuses burden of attacks on victims Threatens viability of businesses postattack Increases pressure for government assistance
Cost to taxpayers	Can lead to higher budget deficits or higher taxes
Cost to future policyholders	May induce some large policyholders to self-insure May force some firms out of business Source of potential cross-subsidies for terrorism risk
Fraction of industry surplus used to compensate for loss following an attack	Threatens ongoing health of insurance industry Can increase cost of capital to insurance industry
Fraction of loss paid by industry	A measure of the role private insurers play in bearing terrorism risk

[13] Insurance industry loss does not include claim payments initially made by insurers but later reimbursed by taxpayers. Similarly, taxpayer cost does not include the amount subsequently recovered from future policyholders.

The fraction of loss that remains uncompensated after an attack is calculated as the sum of uninsured and unpaid insured losses divided by the total loss in the attack. This measure represents the burden that a terrorist attack in the United States places on individual property owners and is related to firms' ability to recover effectively in the aftermath of any terrorist attack.[14] We consider the ratio of uncompensated to total loss rather than the magnitude of uncompensated loss because the former seems more closely related to both the equity and economic recovery issues that this measure aims to represent.

The cost to taxpayers reflects an issue of direct concern to the federal government: the financial burden placed on the public due to a particular intervention in insurance markets. This cost includes both costs directly under the TRIA program and costs due to any postattack compensation for uninsured or unpaid insured loss.

The cost to future policyholders reflects the increased cost of insurance policies in the form of a surcharge placed on commercial property-casualty insurance premiums. Increasing the postattack cost of insurance could reduce the utility of insurance for nonterrorism risks and lead to cross-subsidies between lower- and higher-risk areas.

The fraction of industry surplus used to compensate for losses following an attack reflects the postattack health of insurance markets. The fraction of surplus used, as opposed to the magnitude, seems appropriate, because credit rating agencies (such as A. M. Best or Standard and Poor's) consider this ratio to be a measure of industry health. These rating agencies begin to downgrade firms if losses exceed roughly 10 percent of their surplus. If a terrorist attack consumed too large a fraction of industry surplus,[15] the economy might be affected as insurance became more expensive, not only for terrorism but for auto, property, and a variety of other insurance lines.

The fraction of loss paid by industry reflects the extent of private insurance industry participation in covering terrorism loss and is therefore an important metric of alternative government interventions in the market for terrorism insurance.

Key Uncertainties

The simulation model's estimate of these outcome measures can depend crucially on the values assigned to each of several input parameters. In many cases, as described previously, no empirically based agreement exists on their values or even the proper

---

[14] Firms may also be able to access funds to finance recovery from other sources, such as the bond market, so one might argue that the insurance industry is not critical to recovery. Some research has shown, however, that adequate insurance coverage is important for a business' long-term success of a business following a natural disaster (Alesch et al., 2001).

[15] As discussed later, we base the fraction of surplus used on estimates of the surplus that insurers allocate to the insurance lines covered by TRIA (not the entire surplus of the commercial property-casualty insurance industry).

probability distribution to place over their values. This monograph's RDM analysis thus compares the performance of alternative interventions across a large ensemble of plausible futures, created by considering many combinations of these uncertain model input parameters.

Table 2.3 shows the entire set of uncertain model input parameters that we vary to create the ensemble of plausible futures used for the analysis. Each parameter and the data used to define its uncertain range are discussed in detail in the appendixes. We collect all the uncertain model inputs and their ranges here in a single table to facilitate subsequent discussion of the RDM analysis. In general, the results presented in this monograph do not depend strongly on the choice of end points for these uncertain parameter ranges. Rather, the results depend on the parameter values that cause losses suffered by particular stakeholders to exceed certain values. Thus, the results depend on whether such threshold values are included within the uncertainty ranges we consider.

We focus on the parameters in Table 2.3 because their uncertainty can have a large impact and because we judge them to be particularly important to the comparative performance of the alternative government interventions. These parameters fall into three groups, corresponding to the components of the model they address: preattack industry parameters, which affect take-up rates in response to particular government interventions; loss from attack parameters, which affect the loss generated in any future terrorist attack; and postattack government compensation parameters, which relate to any government decision to compensate losses after any attack.

Relationships in the Model

The simulation model estimates how each policy intervention would perform according to the outcome measures over a wide range of futures as defined by different combinations of the uncertain model-input parameters. As shown by the shaded boxes in Figure 2.1, the simulation model first estimates preattack insurance industry take-up rates for property coverage against conventional and NBCR attacks, contingent on the government intervention. An attack then occurs that generates a loss, some of which insurance would cover. After any attack, the government may decide to provide postattack compensation for uninsured or unpaid insured losses. The uncertainties affect how the model estimates take-up rates, loss from any attack, and the government's postattack compensation decisions. The resulting distribution of losses then defines the outcome measures in each particular scenario. We provide an overview of the components of the model:

- the terrorist-attack model
- the take-up rate model

Table 2.3
Input Parameters Varied to Create the Ensemble of Plausible Futures

Uncertain Input Parameter	Symbol	Possible Values
Preattack insurance industry		
Take-up of commercial property policies for conventional attacks under TRIA (%)	γ_{Conv}	[55, 65]
Take-up of commercial property policies for NBCR attacks under TRIA (%)	γ_{NBCR}	[3, 10]
Exceedance probability[a]	p^e	1%, 0.1%, 0.01%, $500 billion
Demand elasticity	ε_d	[–0.975, –0.325]
Expense factor	h	[1.43, 2.0]
Cost of capital ($)	λ	[0.05, 0.25]
Expected industry copayment beyond $100 billion (%)	c_2^e	[0, 75]
Initial surplus	Q	[$120 billion, $240 billion]
Loss from attack		
Conventional attack types	1-ton bomb, 10-ton bomb	
NBCR attack types	Indoor sarin, radiological bomb, outdoor anthrax, nuclear bomb	
Attack scaling	S	[0.33, 3.0]
Postattack government compensation		
Market retention	R	[$27.5 billion, $100 billion]
Actual industry copayment beyond $100 billion (%)	c_2	[0, 75]
Government compensation for unpaid insured loss (%)	g_v	[0, 75]
Government compensation for uninsured loss (%)	g_u	[0, 75]

[a] The percentage values correspond to an annual loss having the specified exceedance probability. The $500 billion value does not refer to an exceedance probability but rather to the value of the loss.

- the postattack government compensation model
- the insurance-compensation and loss-distribution model.

The Terrorist-Attack Model

The attacks and associated loss estimates we use in our analysis are derived from RMS's Probabilistic Terrorism Model. The RMS model was developed for use in the insurance industry to help property-casualty insurers and policyholders manage their exposure to catastrophic terrorism loss. The model estimates the risk from a wide range of potential terrorist attacks and can be applied at any scale, ranging from an individual building to the entire country. The model uses a probabilistic approach to compute the expected property and WC losses for each attack (whether insured or not), the overall probability of an attack occurring, and the relative likelihoods of different attacks.

Because the RMS model focuses on losses relevant to the commercial property-casualty insurance market, it excludes loss categories that are not normally covered under these insurance lines. Excluded losses include indirect economic losses, damage to government property and injuries to government workers, noncommercial property losses, nonemployee casualties, latent injuries, psychological injuries, and liability losses. While the RMS model does not predict these losses, a property or WC policy may cover them. Appendix A includes details of the RMS terrorism model.

For our purposes, the key characteristics of a terrorist attack include the total insurable loss (hereafter referred to as *total loss*, L), which consists of commercial property loss $\left(L_P\right)$ and WC loss $\left(L_{WC}\right)$, and whether the property loss is covered by insurance policies that exclude NBCR loss.

To efficiently sample a wide variety of plausible terrorist attacks, we begin with six different terrorist attacks: a 1-ton truck bomb, a 10-ton truck bomb, an indoor chemical (sarin) attack, an attack using a radiological device (a dirty bomb), an outdoor anthrax attack, and a 5-kiloton nuclear bomb. Table 2.4 shows the WC and property losses for each attack.

These six attacks present a wide range of total losses, extending from $6 billion to $630 billion (the largest single attack in the RMS model); a wide variation in the relative proportions of property and WC losses (property loss ranges from 30 percent to 80 percent of the total loss); and two conventional and four NBCR attacks.

To capture both the uncertainty in the loss estimates for individual attacks and the potential for multiple attacks, we further extend the range of total loss we model by scaling the RMS loss estimates for each attack up and down by a factor of 3. The RMS model loss estimates for each attack represent the mean values of upward-skewed distributions of the losses the attack would produce whose standard deviations are at least as large as their means.

Figures 2.2 and 2.3 show the property and WC loss estimates for all of the individual conventional and NBCR attacks in the RMS model. These figures also show the range of losses considered in our experimental design as we scale our six cases over the order-of-magnitude range. The range of attacks considered here, with total losses ranging from $2 billion to $1,900 billion, seems to adequately cover the ranges of plausible attacks one can infer from the RMS model.

Table 2.4
Attacks and Loss Ranges

	Loss ($ billions)			
Attack	**WC**	**Commercial Property**	**Total**	**Simulated Range**
Conventional				
1-ton truck bomb	2.1	4.7	6.8	2–21
10-ton truck bomb	11	11	22	7–66
NBCR				
Indoor sarin	4.9	0.67	5.6	2–18
Radiological bomb	0.40	62	63	20–190
Outdoor anthrax	50	70	120	40–360
5-kiloton nuclear bomb	320	310	630	210–1,900

Figure 2.2
Loss Range Considered for Conventional Attacks

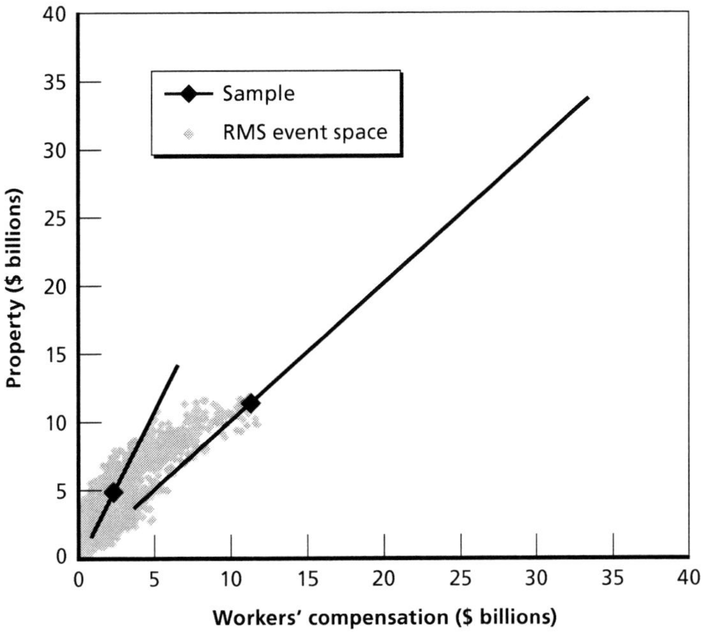

RAND *MG679-2.2*

Figure 2.3
Loss Range Considered for NBCR Attacks

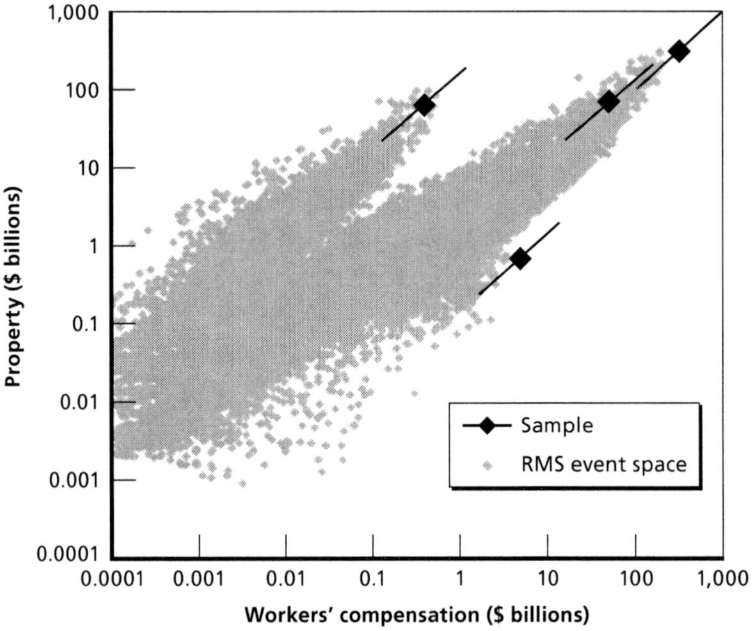

RAND *MG679-2.3*

The Take-Up Rate Model

Prior to any terrorist attack, insurers and their customers decide what fraction of property is covered by terrorism insurance. To determine what part of a loss is insured, we need to estimate policyholder take-up of terrorism insurance. As all WC loss is covered regardless of the source, the take-up rate for WC is always 100 percent.[16] Property policies are often not required to cover losses due to terrorism attacks, however, and the take-up rate of terrorism coverage on property policies is therefore a critical parameter in our analysis.

Our analysis uses a model to predict how commercial property take-up for terrorism insurance would change under different government intervention structures. In particular, we examine how take-up would change if TRIA expired or if Congress enhanced TRIA by requiring insurers to offer coverage for NBCR and conventional attacks. Appendix B describes the model in detail.

General Approach to Modeling Take-Up. Policyholder take-up depends on several factors, including the availability and price of terrorism insurance and the perceived threat of a terrorist attack. Our take-up model starts with an insurer cost model

[16] The term *take-up rate* refers to the proportion of insurance policies that provided coverage for losses—in this case, due to terrorist attacks.

that estimates the costs to insurers of providing terrorism coverage. This cost model uses probabilistic loss estimates from the RMS model, assumptions about insurer behavior and the cost of capital, and the government intervention structure to estimate insurer costs for providing terrorism insurance. Insurers set the price of terrorism insurance based on these costs, and we use an estimate of demand elasticity to calculate the change in policyholder take-up on property policies resulting from changes in the government intervention structure. Our model assumes that insurers can acquire the capital needed to write all the terrorism coverage that policyholders demand (at the prevailing price) for a specified cost per dollar of capital.[17]

The model is calibrated using data on the take-up of terrorism coverage under TRIA and published estimates of the expected take-up of terrorism coverage if TRIA were to expire. The calibration fixes the allowable combinations of parameter values, thus allowing us to use the model to estimate a wide range of scenarios for policyholder take-up rates under the TRIA with NBCR coverage interventions consistent with the literature's estimates of the range of take-up rates with and without the current TRIA program.

Insurer Costs and Policyholder Demand. The RMS model provided an annual exceedance probability distribution, which gave the probability of the total annual loss from terrorism exceeding a given value. This total loss distribution was then decomposed into three components as a function of total loss size: conventional property, NBCR property, and WC. Because many states mandate that loss from fire be covered even if the fire was caused by a peril that is not covered (fire-following coverage), we further decomposed the property loss into fire and nonfire components.

To estimate what fraction of the insurance industry would suffer a loss in a terrorist attack, we used an estimate from Insurance Services Office (2002) of the percentage of the premium base that would be affected by terrorist attacks of different sizes. This estimate ranges from approximately 5 percent for a $10 million loss to a maximum of 70 percent for a loss of more than $100 billion. In other words, in a $10 million attack, only 5 percent of the nation's insurers would have policyholders affected by the attack. Seventy percent the nation's insurers would see claims in a very large attack.

At the individual insurer level, we assume that each insurer pessimistically expects that all losses will occur on its policies, until the proportion of insurers in the industry affected by an attack exceeds the individual insurer's market share.[18] Each insurer then assumes that it will pay claims for losses in accordance with the features (e.g.,

---

[17] Our model is thus not a short-run capacity allocation model in which insurers allocate a set amount of terrorism insurance capacity (that might vary with the government intervention) across policyholders.

[18] For illustration, assume that an insurer has a 7 percent market share. The insurer assumes that its policyholders would suffer all losses in a small attack (one in which ultimately 5 percent of insurers would have policyholders that filed claims). Appendix B contains a discussion of the basis for this assumption and the assumption for larger attacks.

deductibles, copayments, hardness of the cap) of whatever government intervention is in place.

An insurer's total cost of writing terrorism coverage includes the expected loss, including an expense factor, plus the cost of the capital needed to avoid bankruptcy with a given probability. Claim payments may exceed expected losses, and the capital reserve protects the insurer from bankruptcy in such circumstances. The insurer selects a target for the probability that it will go bankrupt due to terrorist attacks, with a lower probability implying that the insurer must obtain more capital.[19]

Given an insurer's cost, the take-up of commercial property terrorism insurance is given by a straightforward supply-demand relationship. Insurers are assumed to set the premium required equal to the cost projected from the cost model, and an elasticity of demand is used to translate the percentage change in cost for property policies to the percentage change in the take-up rate on property policies. Appendix B provides details on this relationship.

Model Calibration. The take-up rate model includes several uncertain parameters. The two key parameters used to calibrate the model are the current take-up under TRIA and the anticipated take-up if TRIA were allowed to expire. Surveys by the U.S. Department of the Treasury (2005) and PWG (2006) have estimated ranges for these values. The model includes several other parameters, some well known and some highly uncertain. These include the elasticity of demand for terrorism property insurance, the expense factor for insurance adjusting and underwriting, the cost of capital, the exceedance probability used in insurer planning, and the insurer responsibility for the insured loss above the program cap. Appendix B contains a discussion of value ranges for these parameters considered in our calibration. To calibrate the model, we predicted the take-up rate without TRIA over a wide range of parameter values. From these runs, we identified those combinations of parameter values that resulted in take-up rates within the range reported in the literature.

Estimating Take-Up for TRIA-with-NBCR-Coverage Interventions. Given the allowable combinations of model parameter values, we then applied the TRIA-with-NBCR-coverage interventions to the take-up rate model and calculated the resulting take-up. As is clear from their descriptions, the key features of these interventions are that policies cover property losses from NBCR attacks, insurer deductibles vary, and insurers' perceptions of the hardness of the program cap vary.

In general, the addition of NBCR losses increases insurer cost and terrorism insurance price and therefore decreases policyholder take-up relative to the current take-up

---

[19] For literature motivating this approach, see Harrington and Niehaus (2004, pp. 82–89, 146–147), who discuss the value of holding capital to reduce the probability of insolvency from claim costs exceeding insurer assets; Dong, Shah, and Wong (1996, p. 213), who present capacity-based pricing formulas that are a linear function of expected loss and the standard deviation or variance of losses; and Cummins (1991, pp. 267–272), who discusses situations in which an insurer will charge expected loss plus a premium to establish a buffer fund needed to avoid the risk of ruin.

under TRIA, while decreasing the deductible or hardening the cap has the opposite effect. Taken together, the features of the TRIA-with-NBCR-coverage interventions result in a take-up rate that can be either higher or lower than the current value for property coverage under TRIA.

The Postattack Government Compensation Model

After an attack, private insurance covers some fraction of the loss, as determined by take-up rates, a cap on losses, and the industry capacity. The government then faces the decision about the extent to which it will compensate any loss that remains uncompensated.

There is a great deal of uncertainty about what fraction of uninsured business property loss (which can include certain business interruption costs) the government would compensate. There are numerous examples of federally funded disaster-assistance programs that provide compensation specifically for uninsured losses. For example, the World Trade Center Business Recovery Grant Program in Lower Manhattan after the 9/11 attacks provided the equivalent of business-interruption and property-damage insurance to small businesses that had not carried insurance at the time of the attacks (Dixon and Stern, 2004, p. 115). The Lower Manhattan Development Corporation administered a federally funded program after 9/11 that covered the costs of restoring the infrastructure of private utilities not covered by insurance or other federal reimbursement (Dixon and Stern, 2004, p. 117). The SBA makes loans to businesses of any size to repair or replace disaster damage to property that the business owns (SBA, 2002). Presumably, a business would not take out such a loan for damage that insurance covered. Even though businesses are required to repay these loans, the loans are subsidized, and they are estimated to cost taxpayers 27 cents per dollar lent (Dixon and Stern, 2004, p. 109).

There are also numerous examples of postdisaster business assistance that is tied indirectly, rather than directly, to uninsured loss. These programs are often targeted at economic recovery after an attack. For example, the World Trade Center Job Retention and Creation Program (funded by HUD) provided payments to large businesses that created new jobs in Lower Manhattan after the 9/11 attacks (Dixon and Stern, 2004, p. 119). Such programs may not provide direct compensation for uninsured losses, but they may still be linked to uninsured losses: The government may be likelier to adopt such programs and in greater numbers when there are higher uninsured losses.

This discussion recognizes a direct or indirect link between uninsured loss and government compensation, but substantial uncertainty exists over the strength of the link.[20] To account for this uncertainty, we allow the government compensation of the

---

20 Studies in other contexts have not found a strong inverse relationship between insurance and federal disaster assistance. For example, Dixon, Clancy, et al. (2006, p. 71) found some empirical evidence that a higher take-up rate for flood insurance by homeowners was associated with less disaster assistance but that the relationship was

uninsured loss to vary between 0 and 75 percent.[21] When government compensation of uninsured losses is zero, any government compensation to businesses after an attack is independent of the amount of uninsured loss.[22]

After an attack that causes a large loss, any insured loss that goes unpaid due to insurer insolvencies would presumably be paid by state-run insurance guarantee funds. Guarantee funds are usually financed by assessments on insurers or policyholders in the state. We assume that insurers, either directly or through guarantee funds, pay the entire insured loss until the industry surplus attributable to TRIA lines is exhausted.[23] For the insured loss beyond this level, the government may feel strongly compelled to help compensate victims whose claims for insurance go unpaid. To account for this contingency, government compensation of the unpaid insured loss varies, in the scenarios we consider, from 0 percent and 75 percent.

The Insurance-Compensation and Loss-Distribution Model

The final step in the simulation modeling is to allocate the loss from terrorist attacks among various parties. To simplify our analysis, we do not estimate policyholder deductibles or coverage limits. The effect of this simplification is to lump deductibles and losses exceeding coverage limits together with the insurance industry's share rather than to include them in the insured's share. As discussed in Carroll et al. (2005), the sum of these loss components is expected to be small compared to the total insurance industry share of losses; therefore, our simplification has little effect on the overall loss distribution.[24]

Here, we summarize how losses are distributed among the different stakeholders under the different government interventions considered in this study. Appendix C provides details of the loss distribution calculations.

---

not large and was statistically significant only for the relatively small amount of disaster assistance that overlaps with the insurance coverage available from the National Flood Insurance Program.

21 It seems plausible that the percentage of uninsured loss compensated would rise with the size of the attack, at least until the attacks became exceedingly large. Data are not available to confirm or reject this hypothesis, and we do not vary the proportion of uninsured loss compensated by attack size.

22 Government disaster outlays that are completely independent of the uninsured loss (for example, driven only by the size of the attack) would presumably be independent of the particular government intervention in terrorism risk insurance markets and thus provide no basis for distinguishing among the interventions considered in this analysis.

23 In our accounting framework, insurers pay assessments levied by guarantee funds. States may allow insurers to pass some of these assessments on to policyholders. In some scenarios, doing so would increase the cost to future policyholders (cost to future policyholders is discussed in the next section) above that shown in our analyses.

24 A large policyholder might have a lower coverage limit for losses due to terrorism than to losses due to other perils. However, this limit is typically for a nationwide portfolio of properties, and the limit may not be reached by a geographically concentrated attack. Thus, it is reasonable to expect that policy limits would not come into play for a wide range of terrorist attacks, at least for large policyholders.

If TRIA is allowed to expire, the government will not be subject to paying for the insured loss through the TRIA program, although the government may end up paying postattack compensation for the uninsured or unpaid insured loss. Some fraction of the insured loss may also go unpaid if the insured loss exceeds the surplus available for paying terrorism claims and the government declines to pay losses. Losses are therefore distributed across four categories: (1) private commercial property and WC insurers (insurance industry); (2) owners of uninsured commercial property and uninsured businesses in those properties (uninsured loss); (3) insured commercial property owners, businesses, and employees not receiving insurance payments (unpaid insured loss); and (4) taxpayers. Table 2.5 summarizes the distribution of losses.

In scenarios with either the current TRIA or one of the NBCR extensions considered here, the insured loss is shared among insurers, future policyholders, and taxpayers. TRIA requires that the government recoup its payments from future policyholders until total outlays by insurers and future policyholders equals the marketplace aggregate retention amount of $27.5 billion. TRIA also provides the option for the government to recoup additional money if it desires. Our analysis allows the retention amount to vary between $27 billion and $100 billion. The law does not specify who is responsible for paying any insured loss above the $100 billion cap; in our analysis, this loss is shared among insurers, taxpayers, and unpaid insured. As shown in Table 2.6, the total loss is thus distributed across five categories.[25]

A key factor influencing the distribution of losses under TRIA is the amount paid by the insurance industry toward the insurer deductible. The aggregate industry payment under TRIA depends on the number and sizes of insurers suffering losses and on the distribution of losses among those insurers. In general, the more widely spread the losses are among insurers, the greater the industry share. As discussed in Appendix C, we assume that the applicable industry deductible is the fraction of the premium base affected by the attack multiplied by the aggregate industrywide deductible ($36 billion

Table 2.5
Distribution of Losses with No Government Terrorism Insurance Program

Loss Category	Amount of Loss Borne
Insurance industry	Lesser of (1) industry surplus or (2) WC loss + insured property loss
Uninsured loss	Property loss not covered by terrorism insurance – postattack government compensation
Unpaid insured loss	Insured loss in excess of industry surplus – postattack government compensation
Taxpayers	Postattack compensation for uninsured and for unpaid insured

[25] Note that, even when insurers initially pay insured losses, we do not assign those payments to insurers that are later reimbursed by taxpayers. Taxpayer payments do not, in turn, include outlays later recovered from future policyholders.

Table 2.6
Distribution of Losses Under TRIA-Based Intervention Structures

Paying Party	Amount Paid
Insurance industry	Lesser of (1) industry surplus or (2) industry deductible (or insured loss, if less) + industry copayment up to TRIA cap (or insured loss, if less) + industry copayment over TRIA cap (if any)
Uninsured loss	Property loss not covered by terrorism insurance – postattack government compensation
Unpaid insured	Insurance industry share (including industry copayment over TRIA cap, if any) in excess of industry surplus – postattack government compensation
Taxpayers	All unrecouped federal payout + postattack compensation
Future policyholders	That part of the federal government payout such that the total paid by insurers plus future policyholders equals a specified retention amount

in 2007). The amount that the industry pays toward the deductible is then simply the lesser of the insured loss or the deductible.

Appendix C also examines a different approach, developed by Kunreuther and Michel-Kerjan (2006), for estimating the insurance industry share. Their approach gives an industry share that is approximately 25 percent lower than our estimate for a $25 billion insured loss and approximately 25 percent higher for a $100 billion insured loss. Our approach and the Kunreuther–Michel-Kerjan approach rely on several assumptions, and it is unclear which is more appropriate.

For the various extensions of TRIA to address NBCR, losses are distributed across the same five categories as with TRIA. However, the allocation will differ from that under TRIA because of changes in the take-up rate, the deductible, and the hardness of the cap.

Experimental Design

To conduct the RDM analysis, we used the simulation model to create an ensemble of plausible scenarios. Each scenario records the performance of a government intervention using the outcome measures, calculated using one combination of the uncertain model input parameters shown in Table 2.3. We aim to choose these scenarios to provide an exhaustive sample of the plausible results from our simulation model, within the constraints of reasonable run times and analysis.

To create the scenarios, we began with a 24,000-point experimental design that combines a 48-point, full-factorial sample over the three uncertain integer parameters (attack type with six values, exceedance probability with four values, and two regimes for belief about TRIA cap) and a 500-point Latin hypercube (LHC) sample over the real-valued inputs.[26] The LHC sample used in this study provides one of the most efficient means of comprehensively sampling a multidimensional space.

We next eliminated those cases in the design for which the preattack industry parameters yield take-up rates with no government program that fall outside the literature's estimates. We calculated the conventional property take-up rate if TRIA were to expire for each point in the design; identified the records for which this no-government-program take-up rate is outside the range predicted in the literature—that is, within 25 percent to 75 percent of the conventional property take-up rates under TRIA; and removed from the design the scenarios with no government program take-up rates outside this range. This process yields 13,704 cases.

We separately analyzed scenarios with conventional and NBCR attacks and, for each of these types of attacks, analyzed scenarios in which the hardness of the existing TRIA cap is uncertain (which we take to be the current situation) and scenarios in which the existing cap is hard. Table 2.7 shows how the 13,704 cases divide among the resulting four designs. The hard cap scenarios can be viewed as a special case of the more general assumption present for the existing cap scenarios: that the cap ranges from hard to quite soft. We base the main part of our analysis on the existing cap scenarios and use the hard cap scenarios to better understand how the hardness of the cap affects the results. Note that more cases yield no government take-up rates consistent with the literature under the assumption that the hardness of the cap is uncertain than are yielded under the assumption that the cap is hard.

We examined the performance of each the six interventions at each of the 13,704 points in the design, for a total of 82,224 cases in the ensemble.

Finally, we considered two ensembles of scenarios: one with the full range of assumptions about postattack government recoupment under TRIA and a sample identical in all the other parameters but with the market retention held in all cases at its minimum value of $27.5 billion. Each ensemble consists of 82,224 cases.

Table 2.7
Numbers of Points in Experimental Designs

Design	Existing Cap	Hard Cap	Total
Conventional attacks	2,838	1,730	4,568
NBCR attacks	5,676	3,460	9,136
Total	8,514	5,190	13,704

[26] In the first regime for beliefs about TRIA, the proportion of insured losses over the $100 billion cap that insurers paid varies from 0 to 75 percent. In the second regime, the cap is hard and insurers pay no insured losses over the $100 billion cap.

The model is written in Microsoft® Excel® and exercised using the CARs™ exploratory modeling software.[27] Each case in the ensemble takes about half a second to run, so that running 82,224 cases consumed about 12 hours of CPU time on the Sony® Vaio® computer used for this analysis.

---

[27] We thank Evolving Logic for making this CARs™ software available for this project.

CHAPTER THREE

Consequences of Allowing TRIA to Expire

In this chapter, we use the methodology described in the Chapter Two to examine the consequences of allowing TRIA to expire. We compare outcomes with and without TRIA in thousands of plausible future scenarios that capture uncertainties in the type and size of attack, take-up rates with and without TRIA, government compensation for the uninsured and unpaid insured loss, and the other variables listed in Table 2.3 in Chapter Two. We consider scenarios with conventional attacks separately from those with NBCR attacks.

In the context of our model, losses caused by terrorist attacks must be borne by insurers, taxpayers, future policyholders, or the owners of the business property subject to the attacks. The first set of results presented describes outcomes with and without TRIA for taxpayers, insurers, and the businesses owners. For these scenarios, we assume that the government recoups only the minimum amount from future policyholders. In these cases, the costs to future policyholders are relatively small and can be ignored. This first set of results suggests that TRIA offers a trade-off between taxpayers and insurers: Under TRIA, the federal government assumes more risk for the largest conventional attacks, and, in return, the insurance industry assumes a larger share of the risk for smaller conventional attacks. In the next part of the chapter, we examine this trade-off in more detail, concluding with findings on expected taxpayer cost with and without TRIA. We then return to future policyholders, analyzing the implications of a government decision to recover more of its outlays from future policyholders than the minimum required by TRIA.

For most of the findings presented in this chapter, we consider scenarios in which the TRIA cap varies from hard to very soft—that is, insurers end up paying anywhere from 0 to 75 percent of the insured losses once the TRIA cap has been reached. The assumption about the softness of the cap does not affect the estimates of take-up rates with and without TRIA because, as discussed in Chapter Two, these estimates are based on empirical data and thus already include whatever perceptions insurers have about whether the cap is soft or hard. Assumptions about the softness of the cap also do not affect our finding on the incidence of losses for conventional attacks, because none of the conventional attacks examined results in losses exceeding the TRIA cap. However, the assumption about the softness of the TRIA cap does affect loss distribution for NBCR attacks resulting in insured losses exceeding $100 billion, and we discuss how the results would change if the cap were hard—that is, if insurers were never required to pay those insured losses exceeding $100 billion.

Take-Up Rates with and Without TRIA

Based on a review of studies on take-up rates with TRIA in place, we set the take-up rate on property policies under TRIA for conventional attacks at between 55 and 65 percent. As discussed in Appendix B, previous studies suggest that the take-up rate would fall by 25 to 75 percent if TRIA were to expire, resulting in take-up somewhere between 14 and 49 percent. (See Aon, 2005; U.S. Department of the Treasury, 2005; Cummins, 2006; and Mortgage Bankers Association, 2004.)

Existing data indicate that the take-up rate on property policies under TRIA for NBCR attacks is very low and ranges from 3 to 10 percent. If TRIA were to expire, we assume that the take-up for property coverage would remain unchanged at this low level. Appendix B details the studies on which these figures for conventional and NBCR take-up rates are based.

WC policies cover loss regardless of cause, so the take-up rate of conventional and NBCR terrorism coverage for WC insurance is 100 percent with or without TRIA.

Distribution of Losses with and Without TRIA

Conventional Attacks

Figure 3.1A displays the outcomes assuming that TRIA is in place for the 2,838 scenarios in our ensemble of futures with conventional attacks and a broad range of assumptions about the hardness of the cap (see Table 2.7 in Chapter Two). Each point on the figure represents an individual scenario and shows the cost to taxpayers (horizontal axis), fraction of loss that remains uncompensated (vertical axis), and fraction of industry surplus used in each specific scenario (shading of points).

The plot shows a wide range of outcomes. The scenarios in the lower right portion of the distribution of points have costs to taxpayers approaching $35 billion, use between 10 and 30 percent of industry surplus, and leave only 5 percent of losses uncompensated. The scenarios in the upper left part of the distribution cost taxpayers little or nothing, use less than 10 percent of industry surplus, and leave roughly 25 percent of losses uncompensated. To put these numbers in perspective, consider a scenario with a preattack take-up rate for property coverage against conventional attacks of 60 percent, an attack with a conventional 10-ton truck bomb detonating in a major U.S. city causing a total loss of $22 billion (roughly the size of the property and WC losses resulting from the September 11 attacks on the World Trade Center), and a postattack

Figure 3.1
Distribution of Losses for Conventional Attacks with and Without TRIA for a Range of Assumptions, Including That About the Hardness of the Existing TRIA Cap

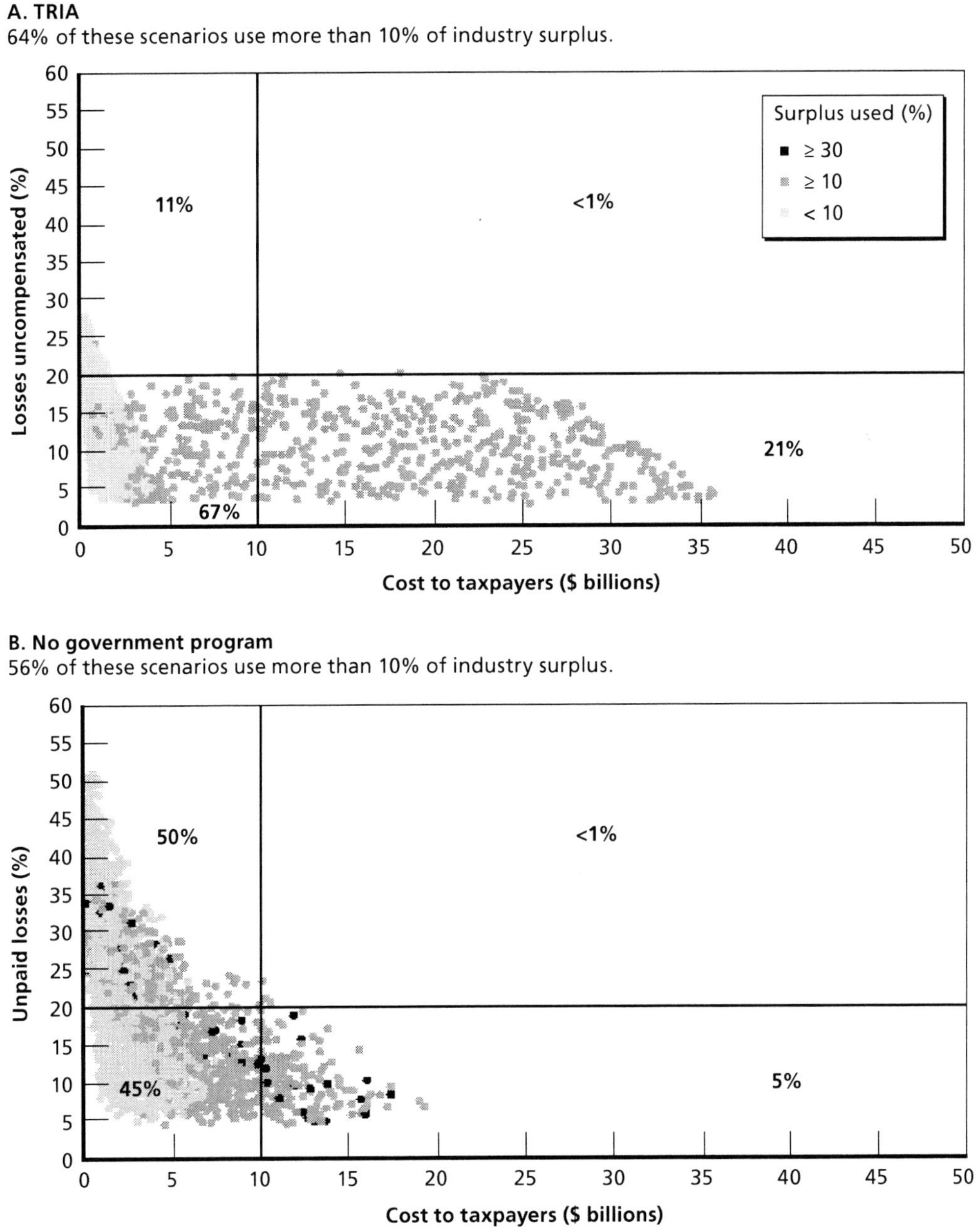

RAND *MG679-3.1*

government decision to compensate 50 percent of the uninsured loss. This scenario would cost taxpayers roughly $2 billion, cost the insurance industry a bit more than

10 percent of surplus ($18 billion), and leave roughly 10 percent of the loss uncompensated ($2 billion). The point representing this scenario would appear in the lower left quadrant of Figure 3.1A and be colored medium gray.

To facilitate comparison with the results for alternative government interventions, it is useful to divide the landscape of Figure 3.1A into four quadrants and count the scenarios that fall into each. It is important to note that these scenario counts have no relationship to the probability of outcomes. These probabilities will be discussed later in this chapter. When interpreting Figure 3.1A and similar figures in this monograph, it is also important to keep in mind that multiple scenarios can plot on top of one another.

The lines in Figure 3.1A mark a taxpayer cost of $10 billion and an uncompensated loss of 20 percent.[1] Twenty-one percent of the scenarios fall in the lower right quadrant with relatively large taxpayer cost and relatively low uncompensated losses; 11 percent fall into the upper left quadrant with relatively large uncompensated loss and low taxpayer cost; 67 percent fall into the lower left quadrant, in which both taxpayer cost and uncompensated loss are relatively low; and less than 1 percent of scenarios fall into the upper right quadrant, in which both are relatively high. We also note that 64 percent of the scenarios use more than 10 percent of industry surplus.[2]

Where a point lies on this landscape depends, of course, on the combination of parameter values that generates that scenario. In general, larger attacks tend to generate scenarios further to the right, and lower levels of postattack government compensation tend to generate scenarios higher up on the plot. As described in Appendix D, we used statistical methods to identify rigorously which of the key uncertain factors in Table 2.3 in Chapter Two have the most significant impacts in shaping the distribution of scenarios shown in Figure 3.1A. In particular, we find that scenarios with costs to taxpayers greater than $10 billion (21 percent of conventional attack scenarios) all have attacks with total losses greater than roughly $40 billion, independent of the value assigned to any of the other uncertain parameters.

We now compare these results to the range of outcomes that could occur if TRIA were to expire. Figure 3.1B shows the outcomes for scenarios with no federal government program in place but otherwise identical to those in Figure 3.1A (e.g., same proportion of uninsured losses compensated by the government, same cost of capital).

---

1 The values for these anchoring points were chosen to allow us to effectively compare the distribution of points with and without TRIA and do not imply that uncompensated loss less than 20 percent or taxpayer cost less than $10 billion are acceptable outcomes.

2 As indicated in Figure 3.1, outcomes for the percentage of surplus used are grouped into three bins: less than 10 percent, greater than or equal to 10 percent and less than 30 percent, and greater than or equal to 30 percent. Ten percent was chosen for the lower cutoff because credit rating agencies, such as A. M. Best and Standard and Poor's, will downgrade an insurer's credit rating if an attack would affect more than 10 percent of surplus. The 30 percent cutoff was chosen so that there were adequate numbers of points in each of the bins (considering both conventional and NBCR attacks) to allow meaningful comparisons.

The key difference between the scenarios with TRIA and those in which TRIA expires is that for the latter take-up is lower and there is no risk-sharing between insurers and government. Allowing TRIA to expire generates an ensemble of scenarios with a smaller range of costs to taxpayers but a wider range of uncompensated losses. Only 5 percent of the scenarios have relatively high taxpayer cost and low uncompensated loss, 50 percent have relatively high uncompensated loss and low taxpayer cost, 45 percent have both relatively low taxpayer cost and uncompensated loss, and 56 percent use more than 10 percent of industry surplus.

As described in Appendix D, we can identify which of the key uncertain factors in Table 2.3 in Chapter Two have the most significant impacts in shaping this distribution of outcomes if TRIA were to expire. In particular, we find that all the scenarios with more than 20 percent uncompensated losses share a common feature: The government chooses after the attack to compensate less than 55 percent of uninsured losses. But in some cases in which the government compensates less than 55 percent of uninsured losses, the value assigned to other parameters can still hold the uncompensated losses below 20 percent. If the government chooses to compensate less than 21 percent of uninsured losses after an attack, the total fraction of uncompensated losses is guaranteed to exceed 20 percent, independent of the value assigned to any of the other uncertain parameters.

Comparing Figures 3.1A and 3.1B suggests that letting TRIA expire shifts risks from the taxpayer to property owners in the form of uncompensated losses. About 40 percent of the scenarios shift from relatively low to relatively high levels of uncompensated loss if TRIA expires (from less than 20 percent uncompensated to more than 20 percent uncompensated), while about 15 percent of the scenarios shift from relatively high to relatively low cost to taxpayers (from more than $10 billion to less than $10 billion). The size of the attack and the level of postattack government compensation largely drive these shifts.

Figures 3.1A and 3.1B focus on the overall distribution of scenarios, but it is important to note that letting TRIA expire does not affect all scenarios equally. Figure 3.2 shows the results of TRIA expiration for three example pairs of scenarios (A, B, and C). In each example, the diamond shows the outcome with TRIA, while the square shows the outcome for a scenario without TRIA that is otherwise identical.[3] Examples A and B show scenarios in which a 10-ton truck bomb attack in New York City causes total losses of $22 billion, roughly the same as those from the assault on the World Trade Center. In example A, the government chooses, after the attack, to compensate 50 percent of uninsured losses. With TRIA, this costs taxpayers $2.2 billion and leaves 8 percent of all losses uncompensated. Without TRIA, the same attack and same postattack government decision about compensation would cost taxpayers

[3] Unlike Figure 3.1, Figure 3.2 does not show the fraction of surplus used in each scenario.

Figure 3.2
How Cost to Taxpayers and the Fraction of Losses Uncompensated Change for Three Example Scenarios When TRIA Is Allowed to Expire

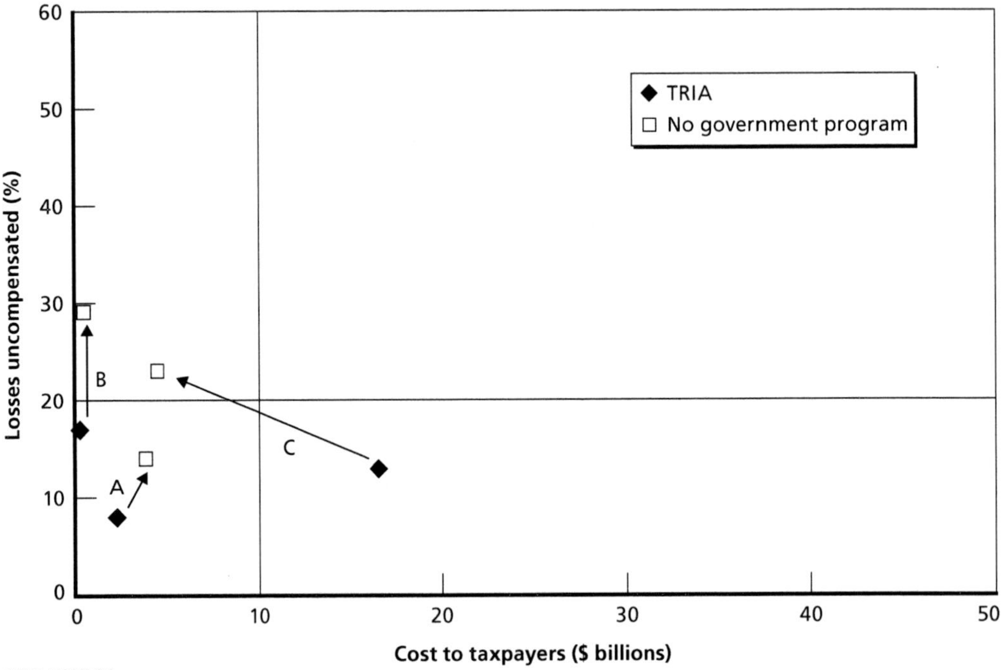

RAND *MG679-3.2*

$3.8 billion and leave 14 percent of all losses uncompensated. Example B shows the results for the same attack but with the government compensating only 5 percent of uninsured losses after the attack. With TRIA, taxpayers would pay $200 million, and 17 percent of losses would be left uncompensated. Without TRIA, the same attack and government decision would cost taxpayers $400 million and leave 29 percent of losses uncompensated.

In examples A and B, both the fraction of uncompensated loss and the taxpayer cost are lower under TRIA, because TRIA boosts policyholder take-up and because TRIA imposes no cost on taxpayers for attacks with losses below the TRIA deductible. The situation changes for larger attacks, in which the government also provides copayments to insurers for losses over the deductible. In example C, one or more simultaneous 10-ton truck bomb attacks in New York City cause $50 billion in total losses and the government decides to compensate 25 percent of the uninsured losses. Under TRIA, taxpayers pay $16.6 billion, funding both copayments under TRIA and the postattack compensation, which leaves 13 percent of losses uncompensated. If TRIA had expired, the same attack and postattack government decision would cost taxpayers $4.4 billion and leave 23 percent of losses uncovered.

The higher take-up rates and absence of government payments for attacks below the TRIA deductible drive the shifts in taxpayer cost and uncompensated loss among the example scenarios in Figure 3.2. As discussed in more detail in the next section, the same two factors explain the shifts among all 2,838 scenarios shown in Figures 3.1A and 3.1B.

NBCR Attacks

Figures 3.3A and 3.3B show outcomes for the 5,676 scenarios involving NBCR attacks with and without TRIA in place. These two scatter plots are similar to those for the conventional attack scenarios, but the scale is much larger. The cost to taxpayers can rise to $1.2 trillion in some scenarios, the fraction of uncompensated losses can approach 100 percent, and the fraction of industry surplus used can exceed 30 percent. The wide spread of points is due, in part, to the tremendous variation in the magnitude of losses that an NBCR event can cause: As discussed in Chapter Two, the losses of the events simulated range from $2 billion to $1.8 trillion.

A comparison of Figures 3.3A and 3.3B suggests that TRIA can make an important difference for the insurance industry in the case of NBCR attacks. The TRIA cap reduces the number of scenarios that use more than 30 percent of industry surplus from 44 percent to 27 percent. The TRIA cap also reduces the scenarios that use more than 90 percent of the industry surplus from 26 percent to 15 percent (not shown).

However, unlike the case for conventional attacks, TRIA has little effect on the fraction of loss that goes uncompensated with NBCR attacks (30 percent or more of losses go uncompensated in 56 percent of scenarios with TRIA compared with 55 percent without TRIA). Uncompensated losses comprise primarily uninsured losses and also include unpaid insured losses. Because the expiration of TRIA has no effect on the take-up rate for property coverage against NBCR attacks (and WC is always covered), allowing TRIA to expire does not change the amount of uninsured losses. Unpaid insured losses under TRIA consist of insured losses that exceed the TRIA cap ($100 billion) and are not covered by postevent government compensation. Unpaid insured losses without TRIA consist of insured losses above the industry surplus ($120 billion to $240 billion) that postevent government compensation does not cover. Insured losses exceed these values only for extremely large events (greater than approximately $230 billion in insured and uninsured losses), so most scenarios have no unpaid insured losses. For scenarios that do, unpaid insured losses are similar in both cases. Thus, the effect of TRIA on total uncompensated losses for NBCR attacks is negligible.

TRIA also does not have a large effect on the percentage of scenarios in which taxpayers pay more than $200 billion. As further investigated in the next section, taxpayer cost is the same or higher under TRIA for NBCR attacks, although the differences are not great relative to the range of taxpayer cost observed over the simulated scenarios.

Figure 3.3
Distribution of Losses for NBCR Attacks with and Without TRIA for a Range of Assumptions, Including That About the Hardness of the Existing TRIA Cap

A. TRIA
27% of these scenarios use more than 30% of industry surplus.

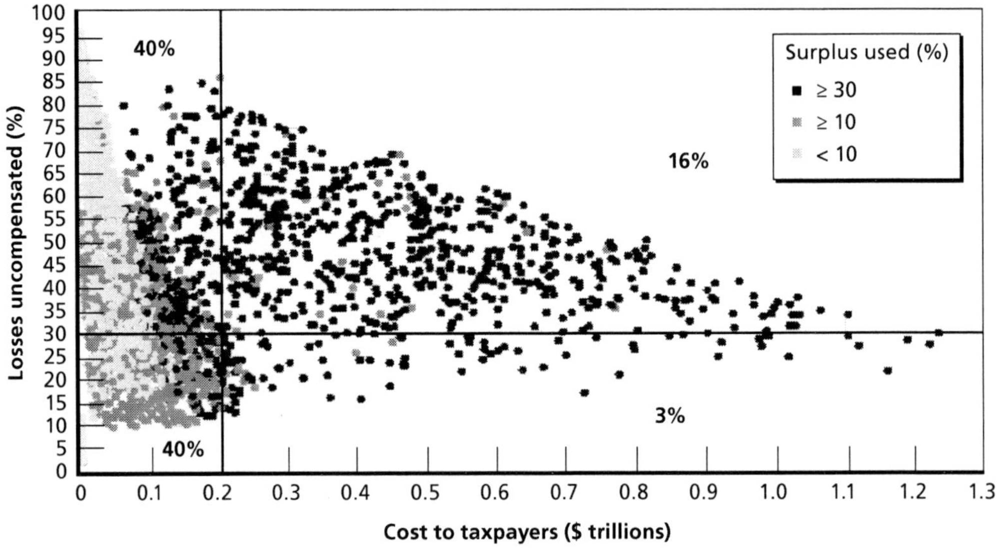

B. No government program
44% of these scenarios use more than 30% of industry surplus.

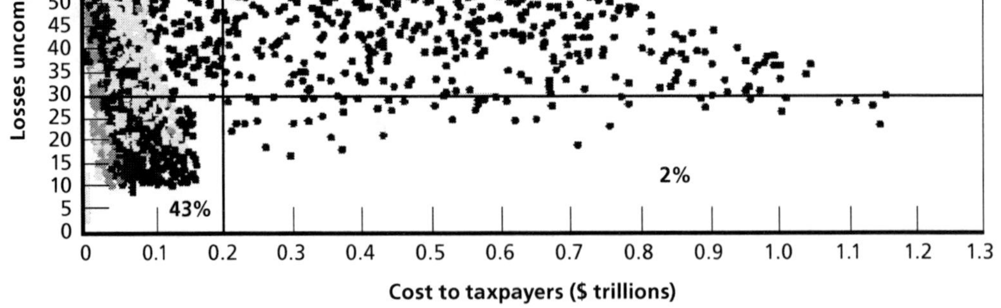

RAND *MG679-3.3*

As a consequence of TRIA's limited effect on uncompensated losses and taxpayer responsibility, the distribution of scenarios among the four quadrants of Figures 3.3A and 3.3B is almost identical with and without TRIA.

The incidence of losses across insurers, taxpayers, and businesses in NBCR attacks depicted in Figure 3.3A is based on an ensemble of scenarios in which the cap can be very soft. If the TRIA cap were hard, insurers' payments for large NBCR attacks would decline, and taxpayers' cost would increase, because insurers would not pay any insured losses in excess of $100 billion, and taxpayers would compensate a proportion (ranging from 0 to 75 percent, depending on the scenario) of the resulting increase in insured but unpaid losses. The fraction of loss unpaid would then also increase for NBCR attacks that produced insured loss over the cap. The impact in Figure 3.3A would be that the percentage of scenarios that use more than 30 percent of industry surplus would decrease, and the points in the figure would tend to shift upward and to the right.

Trade-Offs Between Taxpayers and Insurers

The results shown in Figures 3.1 and 3.3 suggest that TRIA offers a trade-off between the relative amounts that taxpayers and insurers pay for terrorist attacks. Under TRIA, the federal government assumes more risk for the largest conventional terrorist attacks, and, in return, the insurance industry assumes a larger share of the risk for smaller conventional attacks. This section illuminates this trade-off in more detail.

Conventional Attacks

Figure 3.4 compares the cost to taxpayers with and without TRIA for each of the 2,838 scenarios involving conventional attacks. For the scenarios that lie below the 45-degree line in the figure, taxpayers pay more with TRIA in place than without it, while, for the scenarios above the line, taxpayers pay less under TRIA. As described in Appendix D, we find that attack size is the most important uncertainty in determining whether TRIA expiration increases the cost to taxpayers and that, for all scenarios with attacks between $40 billion and $100 billion, TRIA costs taxpayers more. Thus, the plotted scenarios are also distinguished by the size of the attack represented: Dark gray points have losses greater than $40 billion, significantly larger than the losses in the attacks on the World Trade Center.

We find that taxpayers pay less under TRIA than they do without TRIA in most (74 percent) of the scenarios examined. Those scenarios in which taxpayers pay more with TRIA than without generally involve attacks with losses greater than $40 billion. TRIA costs taxpayers less in the smaller conventional attacks because the take-up rate for property coverage against conventional attacks is higher under TRIA. Thus, the fraction of uninsured losses that result from an attack is lower than it is if TRIA

Figure 3.4
Taxpayer Cost for Conventional Attacks with and Without TRIA

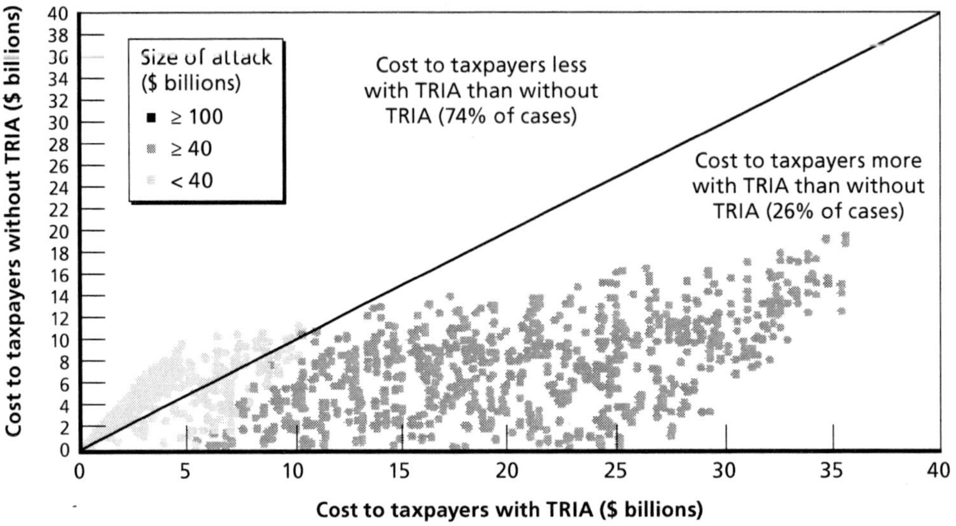

RAND *MG679-3.4*

expires, resulting in lower government compensation for uninsured losses. The government copayment under TRIA begins to impose a cost on taxpayers when insured loss exceeds the mandatory market retention of $27.5 billion. For attacks causing more than $40 billion in damage, this copayment becomes larger than any savings in post-attack government compensation for uninsured losses, causing taxpayer cost to be higher with TRIA than if it expires.

Not only is taxpayer cost lower with TRIA than without TRIA for conventional attacks causing less than about $40 billion in damage, but the role that insurers play in compensating losses in these attacks is greater with TRIA. Figure 3.5 compares the fraction of the total losses that insurers pay under TRIA and if TRIA expires, with the size of the attack again noted. In the scenarios that lay below the 45-degree line, industry pays a larger fraction of the losses with TRIA in place than it does if TRIA expires, while industry pays a smaller fraction under TRIA in the scenarios above the line. The industry share is higher under TRIA for attacks causing less than about $40 billion in damage, because take-up is higher under TRIA and taxpayers make little or no payments for attacks of this size. For the larger attacks, the industry share is less under TRIA than if TRIA expires because of the government copayment for losses over the deductible.[4]

[4] Responsibility for payments over the cap does not come into play for conventional attacks, because none of the attacks examined causes more than $100 billion in loss.

Figure 3.5
Share of Losses Covered by Insurers for Conventional Attacks with and Without TRIA

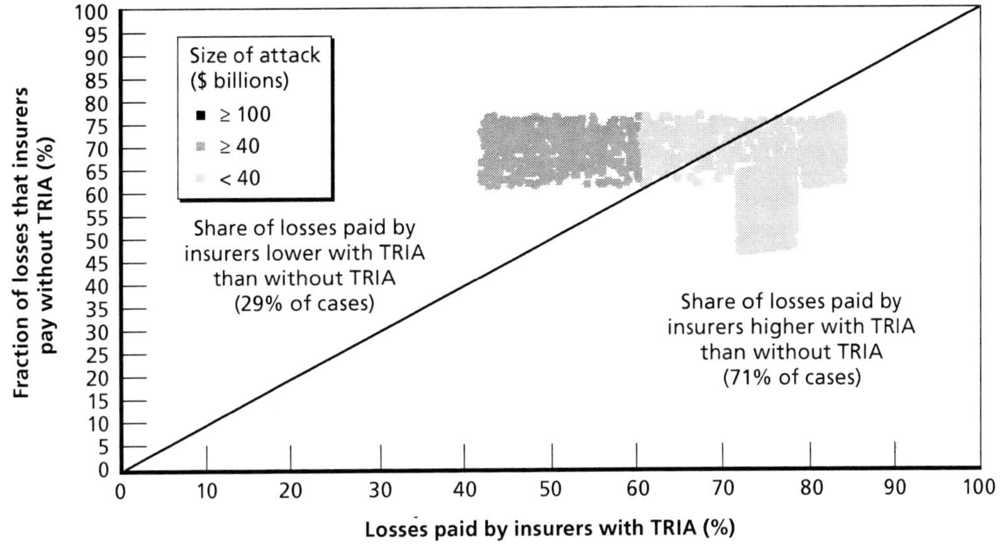

RAND *MG679-3.5*

In summary, when considering conventional attacks, our results indicate that the federal government assumes more risk for the largest terrorist attacks under TRIA than it does without TRIA. In return, the insurance industry assumes a larger share of the risk for smaller conventional attacks under TRIA than it does without TRIA.

NBCR Attacks

TRIA does not offer the same trade-off between large and smaller attacks for NBCR attacks. Figure 3.6 compares the cost to taxpayers under TRIA and if TRIA expires for each of the 5,676 scenarios involving NBCR attacks and a range of assumptions on the hardness of the cap, while Figure 3.7 compares the fraction of total losses that industry pays over the same set of scenarios. In contrast to the outcomes for scenarios with conventional attacks, taxpayer cost for NBCR attacks is never lower with TRIA than without TRIA. This is a result of the fact that TRIA does not affect NBCR take-up and therefore does not reduce government compensation for uninsured losses. For NBCR attacks producing insured losses below the $27.5 billion mandatory market retention, taxpayers make no payments for insured losses, and, because postattack government compensation does not change, overall taxpayer payments are the same with or without TRIA (resulting in scenarios on the 45-degree line). For larger attacks, taxpayers pay for a portion of losses over the market retention, and, in the very large attacks, taxpayers also make higher payments for insured but unpaid losses, to the extent that the TRIA cap limits insurer payments for insured losses claimed once the cap amount has been reached.

Figure 3.6
Taxpayer Cost for NBCR Attacks with and Without TRIA

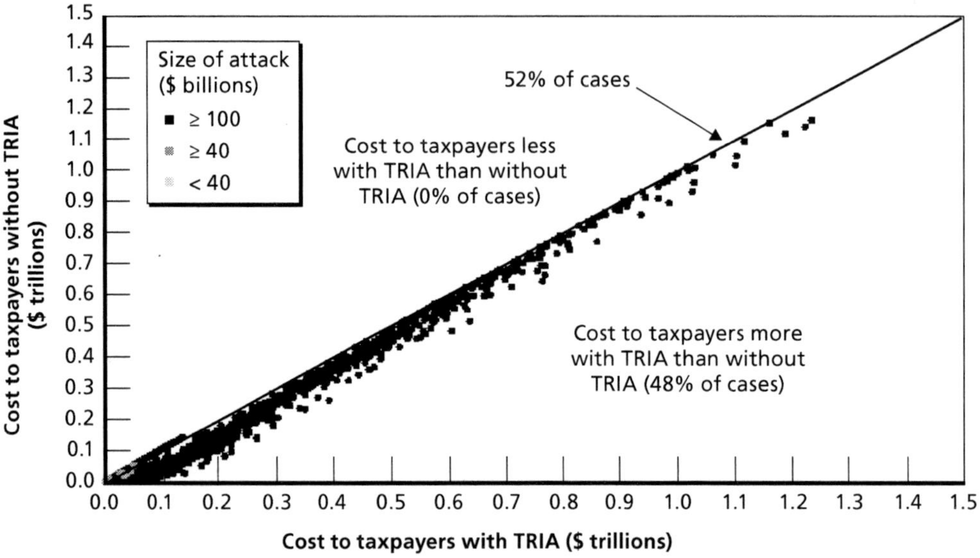

RAND *MG679-3.6*

Figure 3.7
Share of Losses Covered by Insurance for NBCR Attacks with and Without TRIA

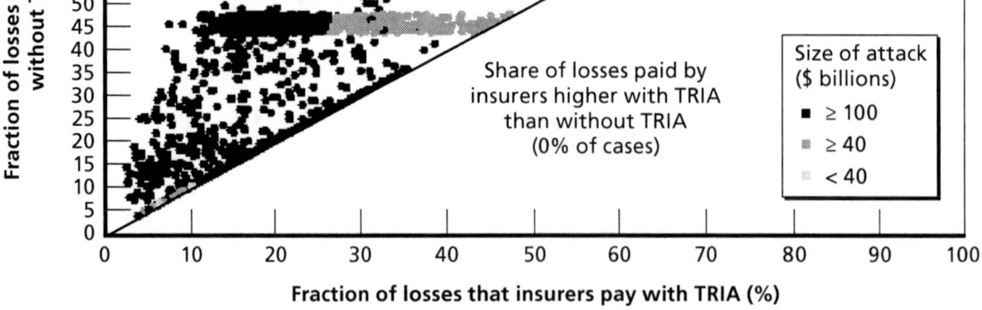

RAND *MG679-3.7*

A Probabilistic Look at the Trade-Offs in Outcomes for Large and Smaller Attacks

Our analysis so far has not considered the relative probabilities of different-sized attacks occurring. We now use existing estimates of the probabilities of large and smaller attacks to compare expected annual taxpayer cost with and without TRIA. The preceding analysis has shown first that taxpayer cost is less with TRIA than it is without TRIA for most conventional attacks with losses less than about $40 billion and is the same with or without TRIA for the smaller NBCR attacks. Second, it has shown that taxpayer cost is higher for the conventional and NBCR attacks with losses exceeding about $40 billion. The relative probabilities of smaller and large attacks can be used to understand the likelihood that expected taxpayer cost under TRIA will be less than or greater than it would be if the program were to expire.

The analysis in Appendix D suggests that attack size is the most important uncertainty affecting the costs to the taxpayers with and without TRIA. The results in this section thus focus on exploring the effects of assumptions about the probability of different-sized attacks. The Appendix D analysis suggests that assumptions about the probability distribution over the other parameters have less effect on the relative costs to the taxpayers with and without TRIA. For simplicity, the calculations in this section thus assume equal weighting over these parameter values (or equivalently such that the parameters are uniformly distributed over the specified range).

As described in Appendix A, RMS uses an expert elicitation process to generate estimates of terrorist attack likelihoods. Estimates from the RMS model suggest that conventional and NBCR attacks with losses greater than $40 billion are each on the order of 1,000 times less likely than are attacks with losses less than $40 billion. If these estimates are correct, then model projects that the expected cost to taxpayers is less under TRIA than it would be if TRIA were to expire, as long as the percentage of uninsured and unpaid insured loss compensated by the government exceeds a very low level.

Figure 3.8 summarizes this result. The vertical axis shows the fraction of uninsured and unpaid insured loss the government chooses to compensate after a terrorist attack, ranging from 0 to 75 percent. The horizontal axis shows the probability of an attack with losses greater than $40 billion relative to its probability in the RMS model, so that a value of 10 on the horizontal axis means that an attack with losses greater than $40 billion is 10 times likelier than the RMS model predicts, and a value of 0.1 on the horizontal axis means that an attack with losses greater than $40 billion is one-tenth as likely as the RMS model predicts.[5] The curve divides the figure into regions in which the expected annual taxpayer cost under TRIA is lower (shaded) and higher

---

[5] For example, horizontal axis values $= \left(P_{>40}\right)_{model} / \left(P_{>40}\right)_{RMS}$, where $P_{>40}$ is the probability of an attack loss greater than $40 billion.

Figure 3.8
Relationship Between Expected Annual Taxpayer Cost with and Without TRIA as the Probability of a Large Attack and Amount of Government Compensation Vary

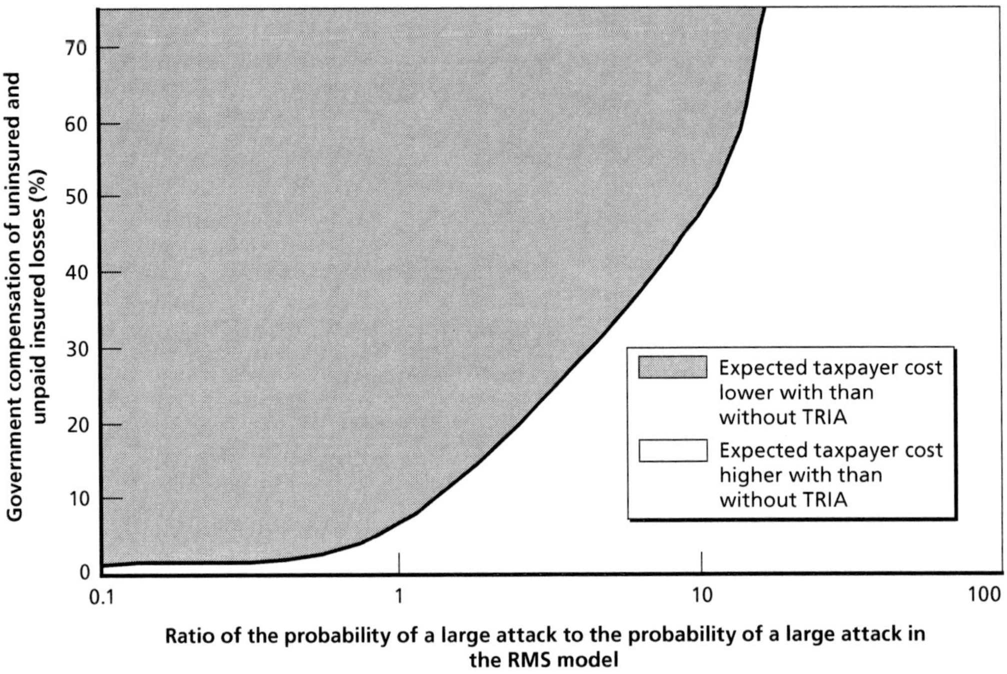

RAND *MG679-3.8*

(unshaded) than if TRIA were to expire. Appendix E presents the numerical differences in taxpayer cost on which this figure is based.[6]

For the baseline attack probabilities predicted by RMS, Figure 3.8 shows that the expected taxpayer cost under TRIA is less than it would be without TRIA, as long as the government compensates more than about 5 percent of uninsured losses and unpaid insured losses.[7] Since any projection of the likelihood of terrorist attacks remains deeply uncertain, Figure 3.8 also considers the effects of a wide range of such estimates. We find that, even if the likelihood of a large attack were 10 times the base-

[6] The difference in expected annual taxpayer cost is not particularly large for most of the parameter values examined. For example, at the baseline attack probabilities and when the government compensates 50 percent of uninsured and unpaid insured losses, expected annual taxpayer cost is $140 million lower with TRIA than it would be if TRIA were allowed to expire.

[7] RMS predicts that the probability of a conventional attack greater than $40 billion is 0.07 percent per year (and thus that the probability of no attack or an attack less than $40 billion is 99.93 percent). The analogous percentages for NBCR attacks are 0.14 percent and 99.86 percent.

line estimate, the expected cost to taxpayers under TRIA is less if postattack government compensation exceeds 50 percent of uninsured and unpaid insured losses.[8]

The analysis of expected taxpayer cost assumes that the TRIA cap could vary from hard to quite soft. As discussed previously, taxpayer cost would be higher for large NBCR attacks if the cap were hard than if it were soft. Thus, if the cap were hard, the expected taxpayer cost with TRIA would be lower than it would be without TRIA for a more restricted set of values for the relative odds of a large attack and the government compensation for uninsured and unpaid insured losses.

Cost to Future Policyholders

So far, we have assumed that the government recoups only the minimum amount of outlays that TRIA requires. Under this assumption, the cost to future policyholders is small relative to insurer and taxpayer outlays. However, TRIA allows the government to recoup a greater amount of its outlays and does not specify an upper limit on how much can be recouped. In this section, we explore the implications of higher recoupment levels.

TRIA requires the government to recoup payments until the sum of insurer outlays and policyholder surcharges equals $27.5 billion (the market retention in 2007). Figure 3.9A shows the costs to taxpayers, industry, and future policyholders as a function of attack size for the conventional attack scenarios. The cost to future policyholders is zero until insurer outlays begin to exceed their deductibles (marked by the kink in the insurer outlay curve that occurs at attacks causing about $27 billion in damage), rises to about $4 billion, and then falls back to zero for attacks that exceed about $60 billion in loss. Policyholder surcharges decrease with attack size, because insurer outlays satisfy more of the required market retention as attack size grows. The cost to future policyholders is never more than about 10 percent of total losses for conventional attacks. Future policyholder cost for NBCR attacks, while it also reaches a maximum of about $4 billion, is even less significant because of the much larger total loss possible for NBCR attacks (Figure 3.9B).

Higher recoupment can significantly reduce the cost to taxpayers and increase the cost to future policyholders. Figure 3.10A shows the cost to taxpayers and future policyholders under TRIA for the scenarios involving conventional attacks and with the minimum market retention. As noted above, the cost to future policyholders never rises above $4 billion. Figure 3.10B shows taxpayer and future policyholder costs for the same scenarios, except that the recoupment varies from the minimum required

---

8 Zero on the vertical axis of Figure 3.8 is equivalent to assuming that government compensation after a terrorist attack is completely independent of the amount of uninsured and unpaid insured losses. If government compensation is indeed independent of uninsured and unpaid insured losses, then expected taxpayer costs with TRIA are always greater than expected taxpayer costs without TRIA.

Figure 3.9
Incidence of Losses Under TRIA, by Size of Attack, with $27.5 Billion Market Retention

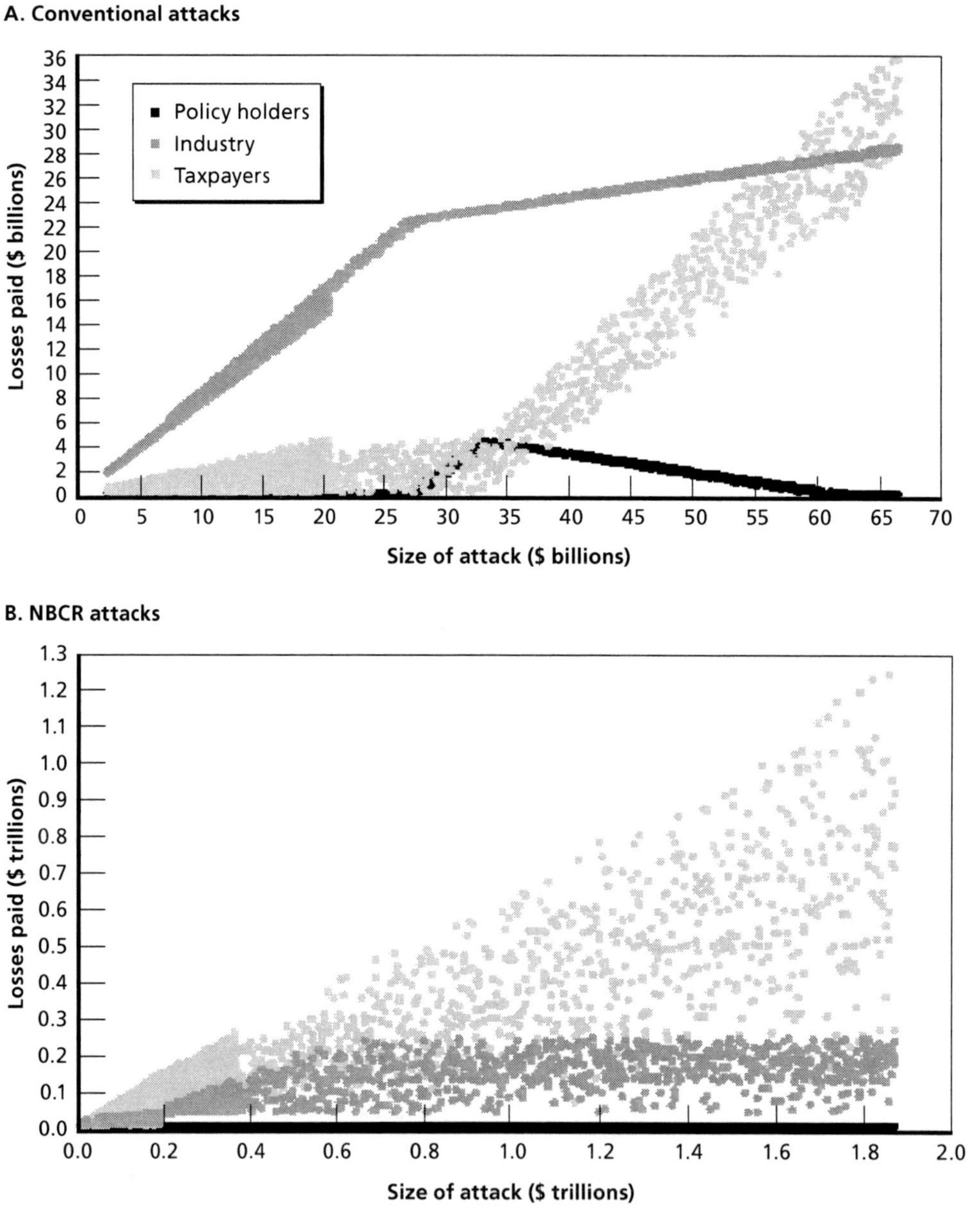

RAND *MG679-3.9*

Figure 3.10
Effect of Higher Recoupment on Taxpayer Cost and Cost to Future Policyholders for Conventional Attacks Under TRIA

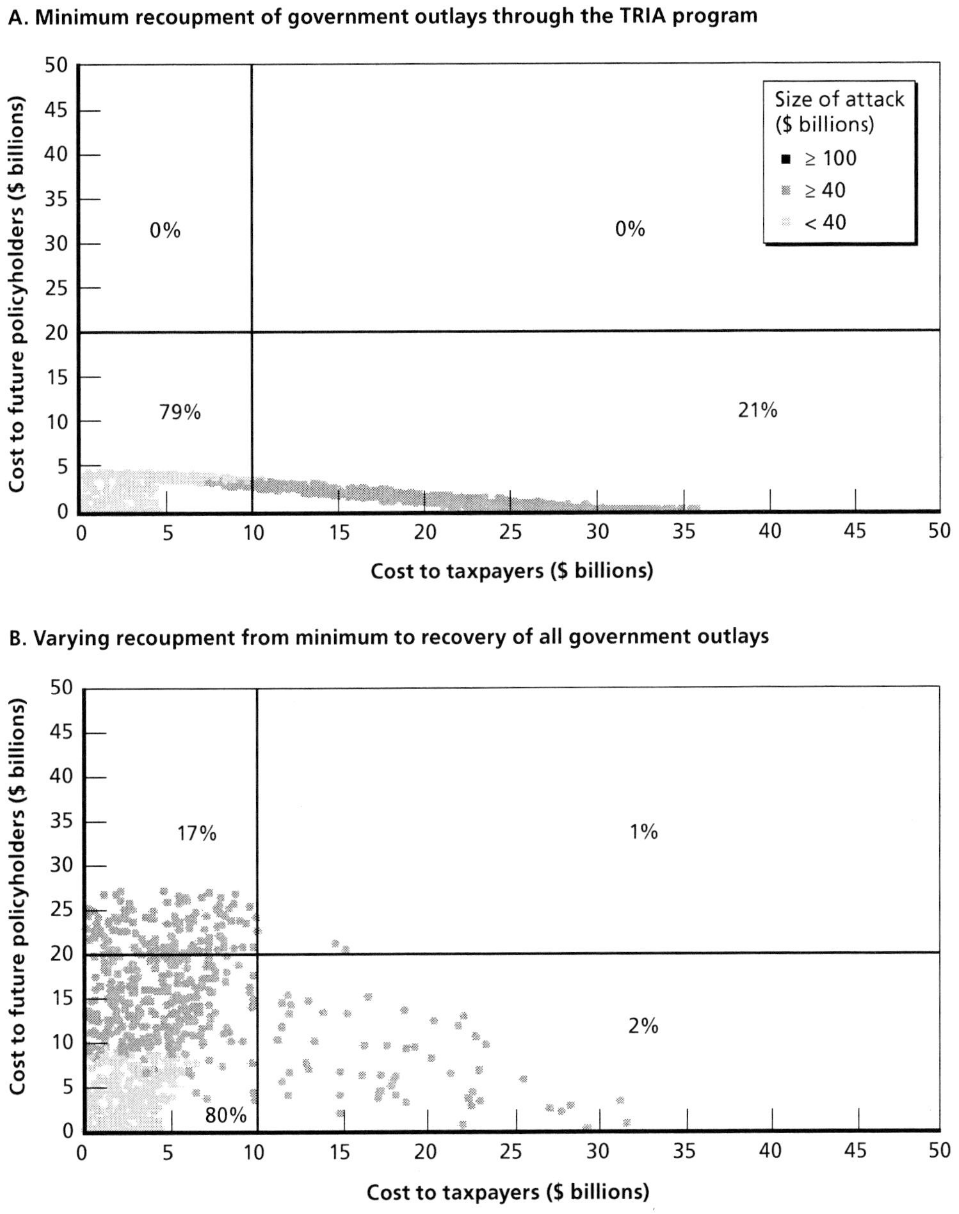

RAND *MG679-3.10*

by TRIA to full recovery of all government payments through the TRIA program.[9] Varying the recoupment over this range significantly reduces the cost to taxpayers and increases cost to future policyholders. Taxpayer payments exceed $10 billion in only 3 percent of the scenarios with higher recoupment, compared with 21 percent when recoupment is limited to the minimum required by law. Policyholder surcharges rise to a maximum of roughly $25 billion, which would take about five years to recover, assuming a 3 percent surcharge on the $180 billion of total premium in TRIA lines.

Higher recoupment thus would expand the range of parameters in Figure 3.8 for which expected annual taxpayer cost is lower with TRIA than it would be if TRIA were allowed to expire. Thus, higher recoupment could make the program more attractive from the point of view of reducing taxpayer liability. In addition, policyholder take-up is independent of the market retention amount in our take-up model,[10] so increasing retention may have no adverse effect on the uninsured loss or industry cost. Higher retention could, however, conceivably reduce coverage for future terrorist attacks if some policyholders decided to reduce or forgo insurance coverage altogether (for terrorism and other risks) due to the increase in insurance costs caused by the surcharge. A higher retention for conventional attacks that results in a 3 percent policy surcharge for five years may also push some firms into insolvency.

Summary

Table 3.1 summarizes the effect of TRIA on the five outcome measures examined relative to no government program for terrorism insurance. Also included in the table is the effect on the take-up rate, which is a key factor affecting of the impact of the intervention. The entries in the table refer to the number of scenarios in which the outcome with TRIA changes as indicated relative to the outcome without TRIA. The number of scenarios with the indicated change should not be associated with a probability of the change occurring. Rather, these summaries give an idea of how TRIA performs over a wide range of futures relative to allowing the program to expire.

Overall, these results suggest that TRIA provides benefits in many dimensions for conventional attacks (see "Conventional Attacks" columns of Table 3.1). Under TRIA, the government assumes much of the risk for very large conventional attacks, which, as shown by the last row in Table 3.1, allows insurers to assume more of the risk for smaller conventional attacks. Considering both government payments under the

[9] Full recovery does not include recovery of payments for uninsured and insured but unpaid loss. For the conventional attacks examined here, the amount recouped never exceeds about $25 billion.

[10] The reason that the amount recouped does not affect take-up rate is that we assume that insurers can pass the policyholder surcharges on to their customers without any effect on the demand for nonterrorism risks. In our model, insurers are thus indifferent between whether payments they receive from the government are financed by taxpayers or by future policyholders when setting rates for terrorism insurance (see Appendix B).

Table 3.1
Summary of Outcomes with TRIA Relative to Outcomes Without TRIA

	Conventional Attacks		NBCR Attacks	
Outcome Measure	**<$40 Billion**	**≥$40 Billion**	**<$40 Billion**	**≥$40 Billion**
Take-up rate	All higher	All higher	All unchanged	All unchanged
Fraction of loss that remains uncompensated after an attack	All lower	All lower	All unchanged	Many unchanged, some higher, some lower
Cost to taxpayers	Most lower	All higher	All unchanged	Most higher
Cost to future policyholders (with minimum recoupment)	Most unchanged, some higher	Most higher	All unchanged	Most unchanged, some higher
Fraction of industry surplus used to compensate loss following an attack	Most higher	All lower	All unchanged	Most unchanged, many lower
Fraction of loss paid by insurance industry	Most higher	All lower	All unchanged	Most unchanged, many lower

NOTE: Share of scenarios with indicated change in outcome. *Some* refers to 1 to 30 percent of scenarios. *Many* refers to 31 to 50 percent of scenarios. *Most* refers to 51 to 99 percent of scenarios.

program and postattack government compensation of uninsured losses, taxpayer cost is lower in most scenarios under TRIA for conventional attacks that result in total losses less than about $40 billion. Conversely, the cost to taxpayers is higher for larger conventional attacks. By increasing the take-up rate for commercial property insurance, TRIA also reduces the fraction of losses that go uncompensated for both smaller and large conventional attacks.

While TRIA provides many benefits for conventional attacks, our analysis suggests that TRIA has only limited benefits for NBCR attacks (see "NBCR Attacks" columns of Table 3.1). Because TRIA has no effect on the property take-up rate for NBCR coverage, the current program has no effect on the proportion of losses that remain uncompensated after NBCR attacks causing less than about $40 billion in losses and, because it limits insurer payouts in very large attacks, increases the fraction of losses that remain uncompensated in most of the scenarios involving large NBCR attacks. TRIA also does not increase the role the insurance industry plays in providing compensation for losses in either large or smaller NBCR attacks (see last row of Table 3.1). In addition, TRIA results in higher taxpayer cost for a substantial fraction of the NBCR scenarios examined.

As shown in the penultimate row of Table 3.1, TRIA does protect the insurance industry to some extent from very large WC claims in the aftermath of a large NBCR attack. Even when the TRIA cap is the softest considered here, insurers still do not pay 25 percent of insured losses once the $100 billion cap is reached, and, if the cap turns out to be harder, then insurer payouts would be lower. This protection from very large losses helps preserve insurers' ability to provide insurance coverage after such an event.

TRIA saves taxpayers money only for conventional attacks that cause less than about $40 billion in losses (see the third row of Table 3.1), but expected taxpayer cost considering all types of attacks is lower with TRIA than it is without TRIA over a wide range of assumptions about the relative probabilities of large and smaller attacks and government compensation for uninsured and unpaid insured losses. Indeed, at current RMS estimates of large and smaller and conventional and NBCR attack probabilities, taxpayers need to compensate only 5 percent of uncompensated losses for expected taxpayer costs to be lower with TRIA than without TRIA.

The results presented in this chapter generally assume that the TRIA cap varies from hard to very soft and that the government would recoup the minimum amount legally required by TRIA from future policyholders. If the government decided to recoup more, the cost to taxpayers would decrease, the cost to future policyholders would increase, and the fraction of loss paid by industry would remain unchanged. If the cap is hard, the cost to taxpayers would increase for large NBCR attacks due to the increase in government compensation for uninsured losses, and the fraction of loss uncompensated would increase as well.

CHAPTER FOUR

Consequences of Requiring Insurers to Offer Terrorism Coverage for Conventional and NBCR Attacks

As noted in Chapter Three, TRIA has done little to increase insurance coverage for commercial property losses due to NBCR attacks. In this chapter, we examine enhancements to TRIA that aim to address this shortcoming by requiring insurers to offer policies that cover terrorism losses due to both NBCR and conventional attacks. We consider the following four variants of such TRIA enhancements that differ in their deductibles and the hardnesses of their program caps:

- The TRIA cap remains unchanged and the deductible remains at 20 percent.
- The TRIA cap remains unchanged and the deductible falls to 7.5 percent.
- The TRIA cap is hardened and the deductible remains at 20 percent.
- The TRIA cap is hardened and the deductible falls to 7.5 percent.

These versions of TRIA with NBCR coverage require insurers to offer bundled conventional and NBCR coverage to commercial policyholders and for policyholders to accept or reject the combined coverage. If policyholders reject the coverage, we do not allow insurers to come back with an offer for only conventional or only NBCR coverage. Recall that we harden the cap by requiring taxpayers to pay insured losses between $100 billion and $650 billion.

We begin our analysis of the consequences of requiring insurers to offer both conventional and NBCR coverage by examining the impact on take-up rate of the enhanced policy. The take-up rate model discussed in Chapter Two and detailed in Appendix B is used to project the take-up rate across a wide range of plausible futures for each of the four variants of the program. As we did in Chapter Three, we then characterize the distribution of losses across taxpayers, insurers, and businesses subject to attack. Findings are presented for large numbers of conventional and NBCR attack scenarios.

Most of the analyses in this chapter assume that the hardness of the existing TRIA cap is uncertain and thus that insurers may end up covering anywhere from 0 to 75 percent of insured losses beyond the $100 billion cap. To better understand how the hardness of the existing TRIA cap affects our results, we also examine the effect of requiring insurers to offer both conventional and NBCR coverage under the special case in which the cap is already hard (insurers have no liability for insured losses beyond $100 billion) in the final section of this chapter. Throughout this chapter, we also assume that the government recoups only the minimum amount of outlays required by TRIA.

Take-Up Rate for a Policy That Covers Conventional and NBCR Attacks

The take-up rate for a policy that provides property coverage for both conventional and NBCR attacks varies a great deal depending on whether the cap on insurer liability is hardened and on the size of the insurer deductible. Figure 4.1 reproduces the take-up rates for conventional and NBCR attacks under TRIA discussed in Chapter Three and shows the range of take-up rates predicted for each of the four variants of TRIA with NBCR coverage considered. The range of take-up rates reflects the variation in the uncertain input parameters in Table 2.3 in Chapter Two across the scenarios examined. As described in Chapter Two, the combinations of model input parameters that generate this range of take-up rates for TRIA with NBCR coverage all also generate ranges of take-up rates with TRIA and without TRIA consistent with estimates in the literature.

The third interval from the top of the figure shows that the take-up rate on property policies can plummet if insurers are required to offer a policy that covers all terrorist attacks and there are no changes in the TRIA cap or deductible. Requiring insurers to bundle coverage for conventional and NBCR attacks causes the cost and, consequently, the price of terrorism coverage to rise, leading to a fall in the take-up rate from what is observed under TRIA.[1] The largest declines occur when insurers perceive the existing cap to be very soft (e.g., they must pay close to 75 percent of insured losses beyond the $100 billion cap). In such circumstances, the price increase required by the increased exposure to losses may be so large that the take-up rate falls to zero.

Lowering the deductible without hardening the cap causes similar declines in the take-up rate: As shown in Figure 4.1, the predictions range from 0 to 63 percent with a 7.5 percent deductible and existing cap, versus 0 to 56 percent when the deductible and cap remain unchanged. The spread in take-up rate does not change much, because the percentage reduction in the cost of offering a policy caused by lowering the deductible is not large in scenarios in which insurers remain liable for a substantial fraction of losses beyond the $100 billion cap.

Hardening the cap but leaving the deductible unchanged eliminates the largest declines in take-up rate observed for the previous two NBCR offer options and can result in a substantial increase in take-up rate. As shown in Figure 4.1, the take-up rate

---

[1] Our analysis assumes that adding NBCR coverage does not shift the demand curve for conventional coverage. We argue in Appendix B that any upward shift in the demand curve that does occur would not be large relative to the increase in the cost of providing the additional coverage and thus that ignoring the increase in demand will not have a large effect on the predicted take-up rate.

Figure 4.1
Property Coverage Take-Up Rates for TRIA and TRIA with NBCR-Attack Coverage When the Hardness of the Existing TRIA Cap Is Uncertain

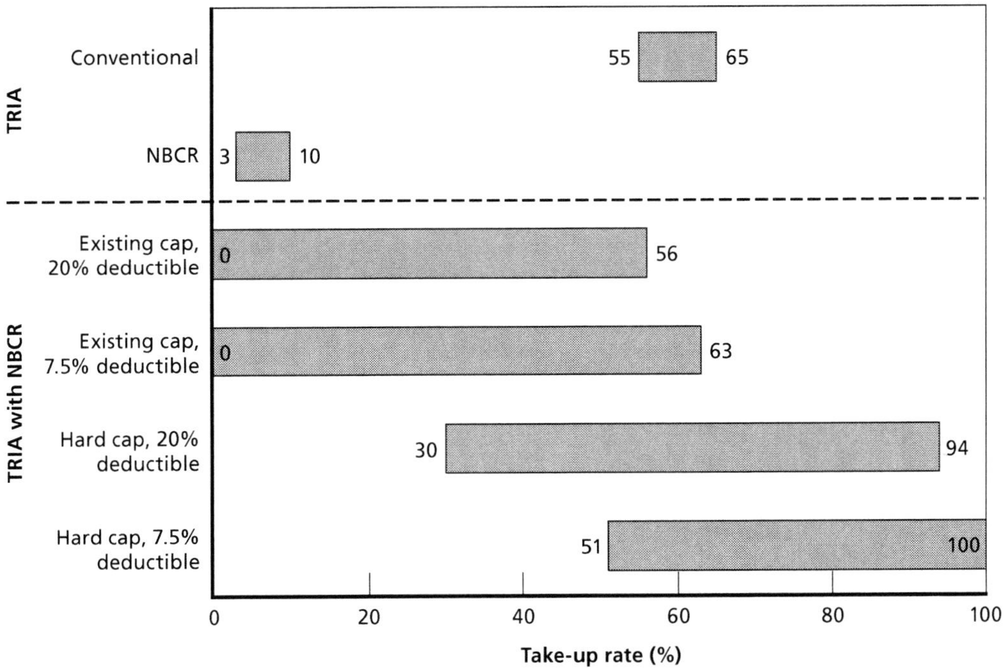

RAND *MG679-4.1*

with a hard cap and a 20 percent deductible ranges from 30 to 94 percent. Hardening the cap offsets some of the additional cost that insurers face when including NBCR attacks in their coverage. The largest declines (down to 30 percent) occur when the existing cap is already fairly hard and thus hardening the cap does not make a great deal of difference. The largest increases (up to 94 percent) occur when the existing cap is very soft and hardening the cap results in considerable cost savings to insurers.

Hardening the cap and lowering the deductible eliminates most of the downside risk of requiring mandatory offer of a more comprehensive terrorism policy, at least as far as the take-up rate is concerned. The bottom interval in Figure 4.1 shows the take-up rate with a hard cap and a 7.5 percent deductible. With these changes in the TRIA program, the range for the predicted take-up rate does not fall much below the take-up rate for conventional attacks under TRIA.

Distribution of Losses for TRIA with NBCR-Attack Coverage

Conventional Attacks

Figures 4.2A through 4.2D display the losses that businesses, taxpayers, and insurers incur in scenarios with conventional attacks for each of the four variants of TRIA with NBCR coverage. These figures plot scenarios in which the existing TRIA cap ranges from hard to very soft. The results indicate that requiring policies to cover both conventional and NBCR attacks without other changes in the program is comparable to letting TRIA expire, in terms of the fraction of losses uncompensated. More than 20 percent of losses go uncompensated in 52 percent of the scenarios examined when the cap and deductible remain unchanged (Figure 4.2A). Recall that, if TRIA is allowed to expire, more than 20 percent of losses go uncompensated in 50 percent of the scenarios (see Figure 3.1B in Chapter Three). Lowering the deductible without hardening the cap (Figure 4.2B) reduces the percentage of scenarios in which more than 20 percent of losses go uncompensated to 42 percent, but the percentage still greatly exceeds the percentage for TRIA (11 percent).

At the same time that it degrades the performance of the program for conventional attacks in terms of uncompensated losses, expanding TRIA to require coverage of both conventional and NBCR attacks without hardening the cap retains some of the downsides of TRIA in terms of taxpayer cost. The percentage of scenarios in which taxpayer cost exceeds $10 billion when the cap is unchanged (17 percent in Figure 4.2A and 18 percent in Figure 4.2B) is closer to the 21 percent found for TRIA than to the 5 percent that results if TRIA is allowed to expire.

When the cap is hardened, outcomes for conventional attacks look much more like those observed under the current TRIA. As illustrated in Figures 4.2C and 4.2D, the percentage of scenarios with more than 20 percent uncompensated losses drops below 20 percent, compared to 11 percent under TRIA. In addition, the percentage of scenarios in which taxpayer cost is greater than $10 billion is comparable to TRIA.

Reducing the deductible and hardening the cap reduces uncompensated losses but increases taxpayer cost to some extent. Reducing the deductible also significantly limits risk to insurer surplus: More than 10 percent of surplus is used in 62 percent of scenarios when the deductible is 20 percent versus 7 percent when the deductible falls to 7.5 percent. Improvements in these dimensions are in part due to increased costs borne by future policyholders, a topic to which will we return later in this chapter.

NBCR Attacks

Even though the take-up rate on property policies for NBCR attacks can rise above the low level observed under TRIA when terrorism policies are required to cover both conventional and NBCR attacks, the take-up rate may also remain very low if the TRIA cap is unchanged. As a consequence, outcomes for NBCR attacks without hardening the cap are not much better than they are under TRIA. The shape of the point clouds in Figures 4.3A and 4.3B do not differ a great deal from the distribution under TRIA (see Figure 3.3A in Chapter Three). The percentage of scenarios in which more than 30 percent of industry surplus is depleted is also comparable.

Outcomes are much different when the TRIA cap is hardened. As shown in Figure 4.3C, more than 30 percent of losses go uncompensated in only 15 percent of scenarios, versus 56 percent under TRIA. The decline in uncompensated losses is due both to the higher take-up rate for NBCR coverage (30 to 94 percent versus the 3 to 10 percent under TRIA) and government payment of all insured losses between $100 billion and $650 billion.[2]

Even with a hard cap, some scenarios still have substantial uncompensated losses. This result is due to the low take-up rate for NBCR attacks in some cases (take-up rates can drop as low as 30 percent) and the partial payment of insured losses once the $650 billion cap is reached. Partial payment of insured losses beyond the $650 billion cap is reflected in the reappearance of scenarios in which more than 30 percent of losses are uncompensated once taxpayer cost reaches roughly $600 billon. When government payments for insured losses exceed roughly $600 billion,[3] insured losses are only partially compensated, and the percentage of losses uncompensated rises.

As expected, hardening the cap by directing the government to pay insured losses between $100 billion and $650 billion does increase the cost to taxpayers in many of the scenarios examined. As shown in Figure 4.3C, taxpayers pay more than $200 billion in 30 percent of NBCR attack scenarios, higher than the 19 percent under TRIA, and the spread of scenarios extends further to the right in the figure than under TRIA. As we will discover, however, taxpayer cost does decline in a substantial number of NBCR-attack scenarios relative to TRIA when TRIA is expanded to require insurers to offer policies that cover both conventional and NBCR attacks.

Lowering the deductible in addition to hardening the cap reduces slightly the fraction of losses uncompensated and slightly increases taxpayer cost (compare Figures 4.3C and 4.3D).

Cost to Future Policyholders

The discussion so far has not considered the cost to future policyholders when TRIA is extended to require offering policies that cover both conventional and NBCR attacks. It is clear that, if the deductible remained unchanged, the maximum cost to future

---

2 Not all of the insured loss is paid under TRIA, because insurers can pay insured losses beyond $100 billion only until industry surplus is exhausted (surplus ranges from $120 billion to $240 billion in the scenarios examined), and the government pays only a fraction of the remaining uninsured loss (0 to 75 percent).

3 By the time the insured loss reaches $650 billion, taxpayer cost will total approximately $600 billion due to (1) approximately $55 billion in outlays for the 85 percent government copayment on the loss between the deductible (roughly $36 billion) and $100 billion and (2) $550 billion in outlays for the insured loss from $100 billion to $650 billion.

Figure 4.2
Distribution of Losses for Conventional Attacks Under TRIA with NCBR-Attack Coverage for a Range of Assumptions, Including That About the Hardness of the Existing Cap

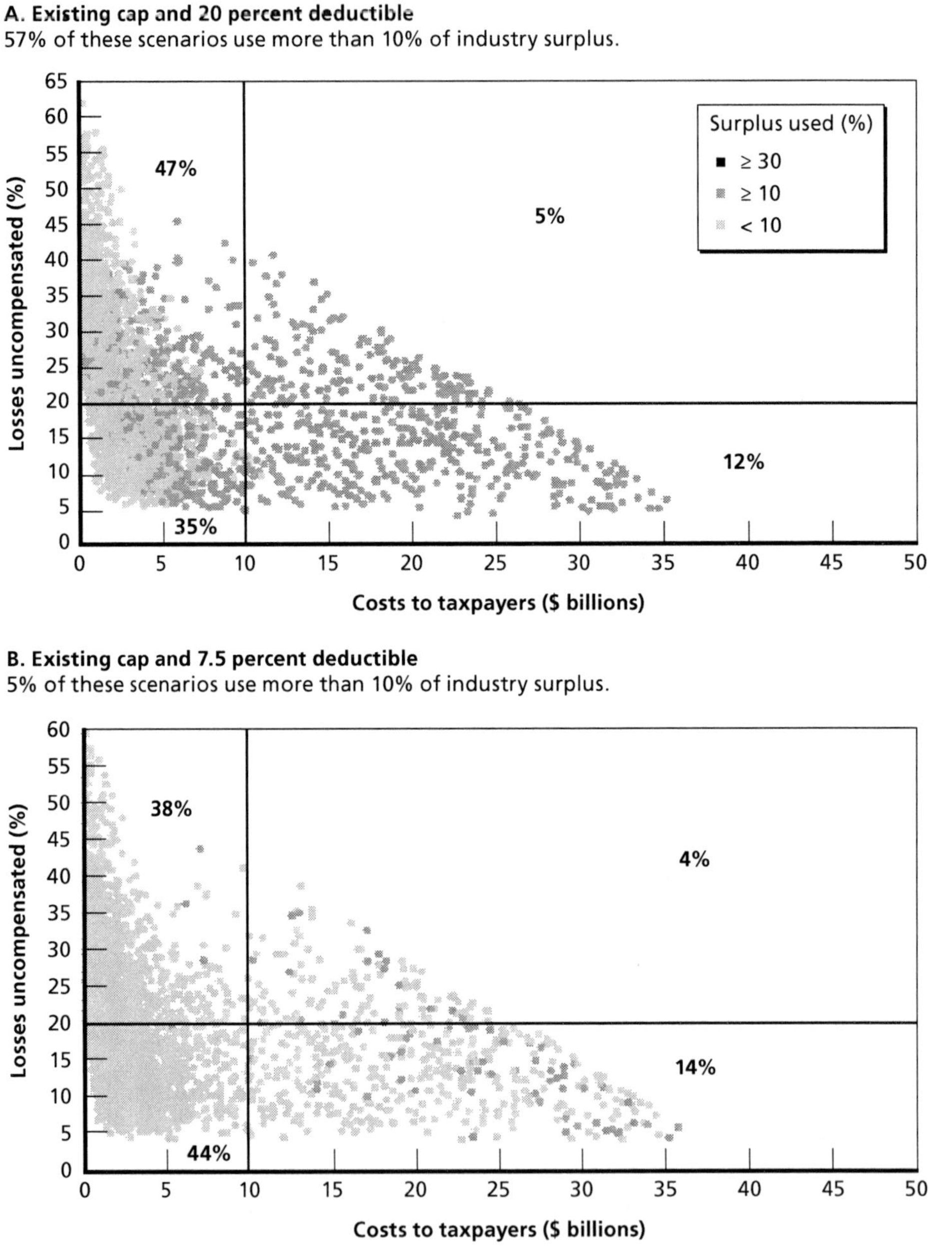

RAND *MG679-4.2*

policyholders would remain similar to that found under TRIA. However, a lower deductible would increase the cost to future policyholders. In this subsection, we

Figure 4.2—Continued

C. Hard cap and 20 percent deductible
62% of these scenarios use more than 10% of industry surplus.

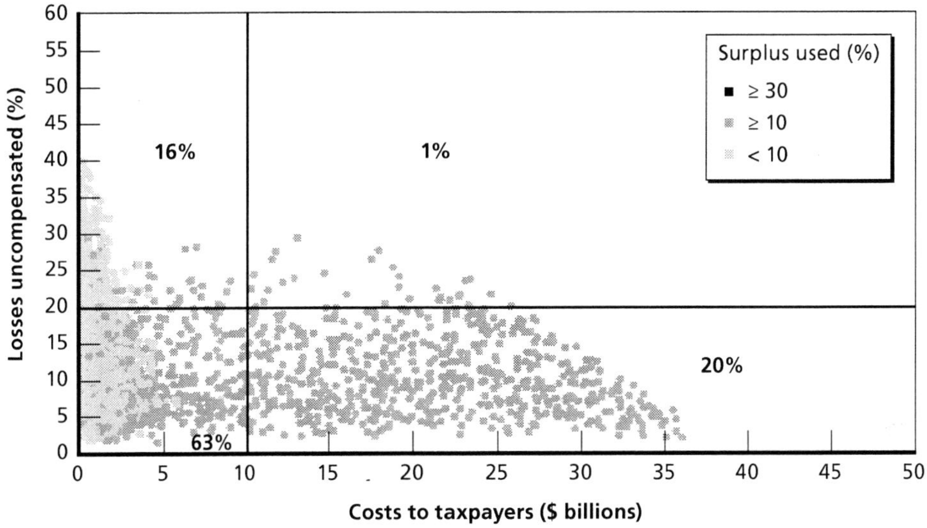

D. Hard cap and 7.5 percent deductible
7% of these scenarios use more than 10% of industry surplus.

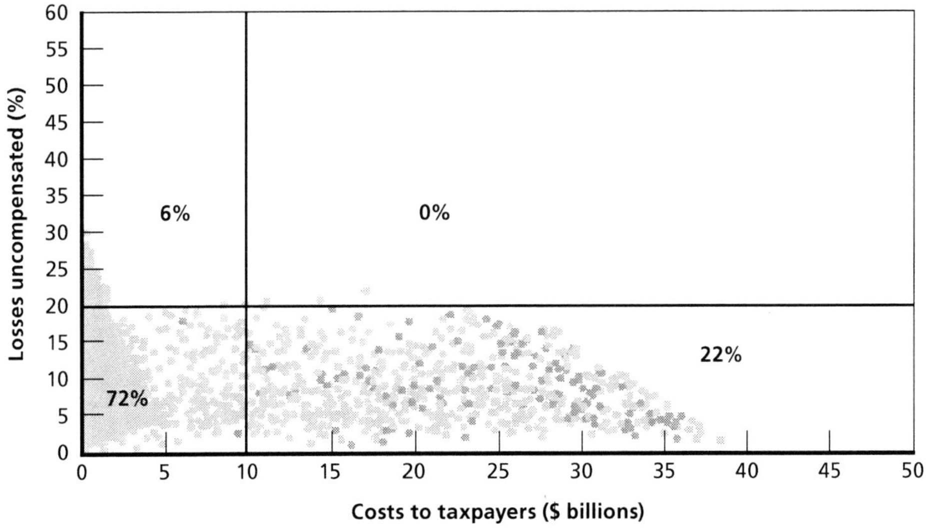

RAND *MG679-4.2 (continued)*

examine this effect. We focus our attention on the version of TRIA with NBCR coverage in which the cap is also hard, because versions that leave the cap unchanged produce less desirable outcomes in terms of take-up rate or fraction of losses compensated and, as discussed later in this chapter, are the robust option when uncertainty about cap hardness is considered.

Figure 4.3
Distribution of Losses for NBCR Attacks Under TRIA with NBCR-Attack Coverage for a Range of Assumptions, Including That About the Hardness of the Existing TRIA Cap

A. Existing cap and 20 percent deductible
31% of these scenarios use more than 30% of industry surplus.

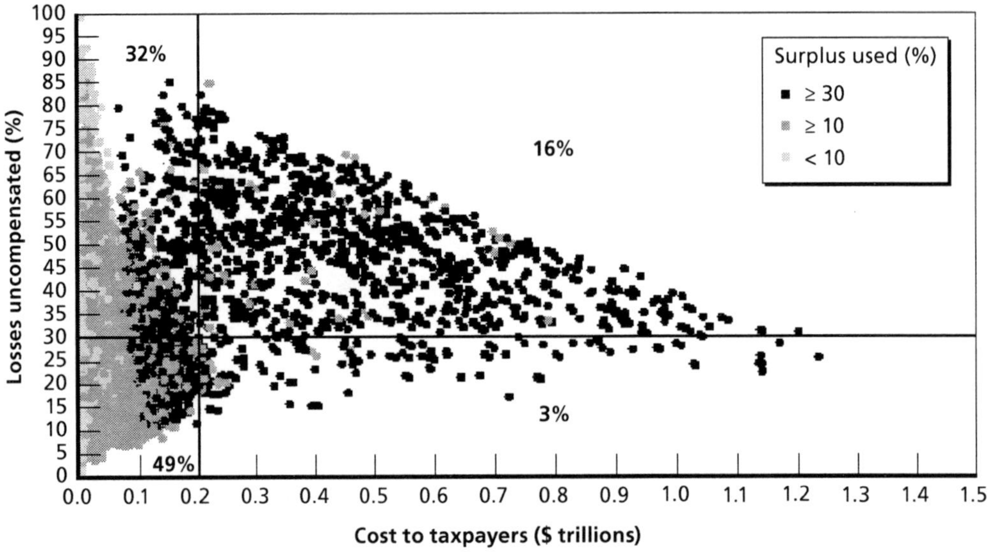

B. Existing cap and 7.5 percent deductible
28% of these scenarios use more than 30% of industry surplus.

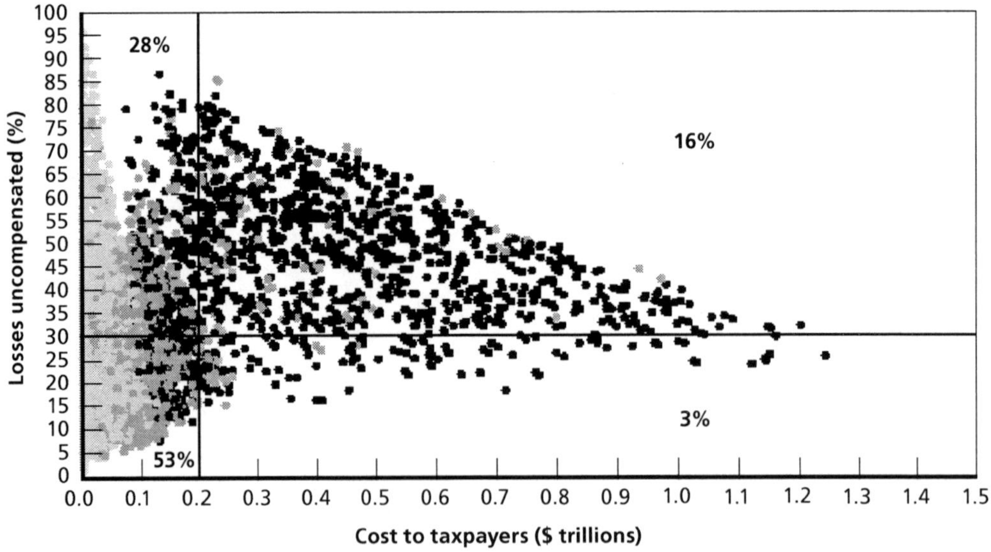

RAND *MG679-4.3*

Figure 4.3—Continued

C. Hard cap and 20 percent deductible
1% of these scenarios use more than 30% of industry surplus.

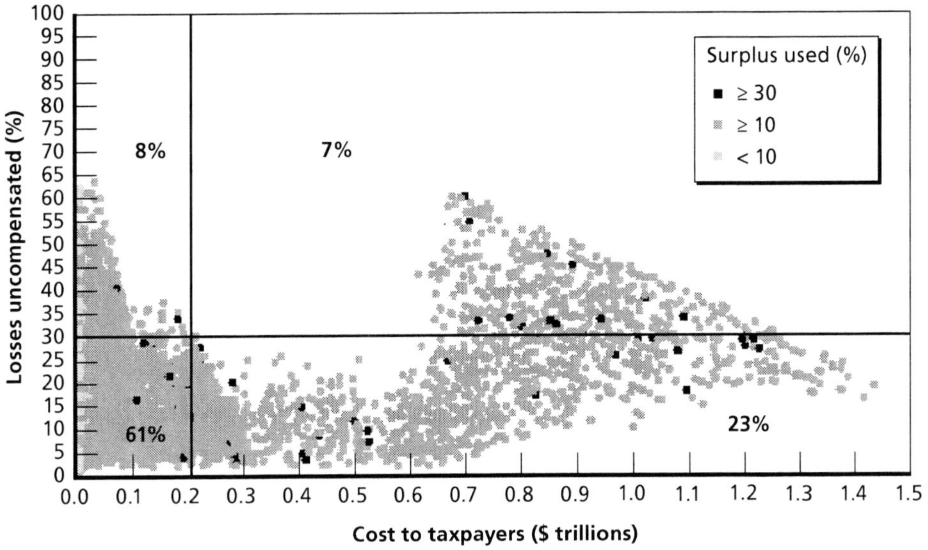

D. Hard cap and 7.5 percent deductible
Less than 1% of these scenarios use more than 30% of industry surplus.

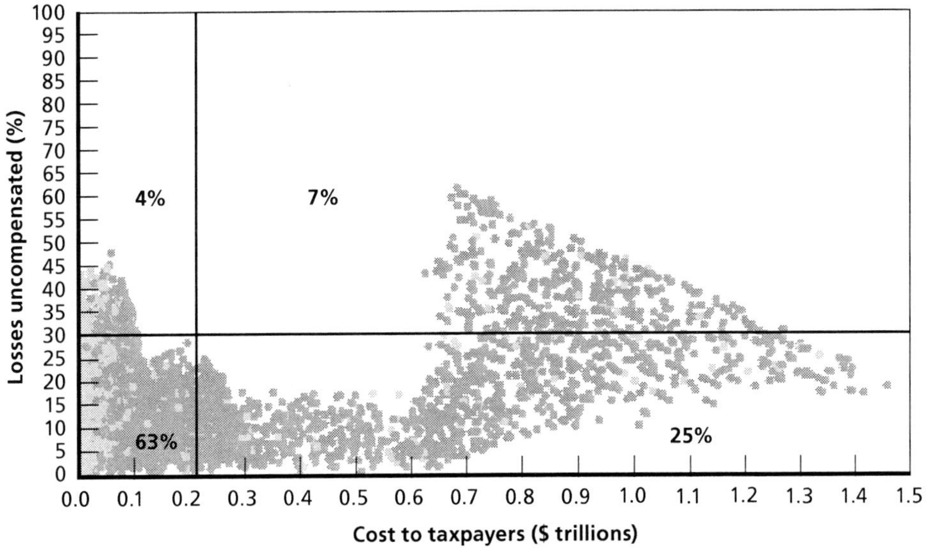

RAND *MG679-4.3 (continued)*

Figure 4.4A shows that cost to future policyholders peaks at $17 billion for conventional attacks when there is a hard cap and a 7.5 percent deductible, considerably higher than the $5 billion under TRIA (see Figure 3.9A in Chapter Three). The cost to

Figure 4.4
Distribution of Losses Under TRIA with NBCR-Attack Coverage, Hard Cap, and 7.5 Percent Deductible

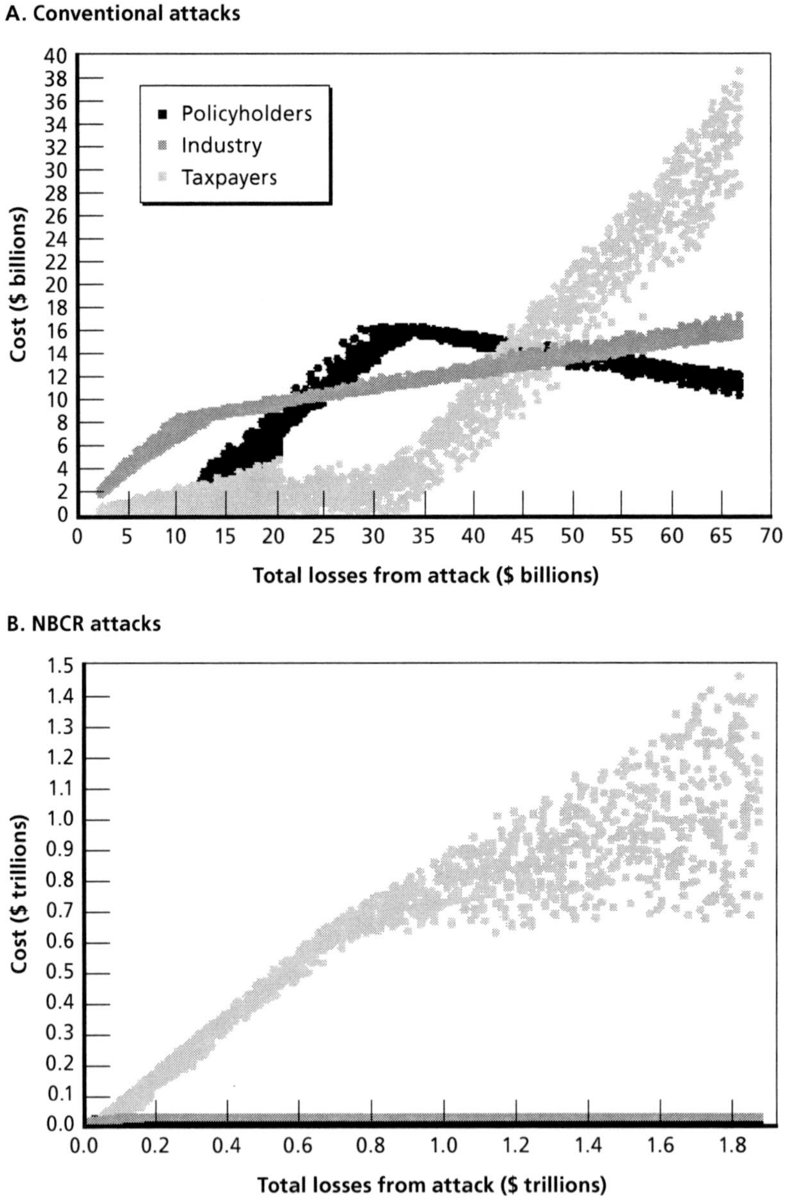

RAND *MG679-4.4*

taxpayers, on the other hand, remains very similar to that under TRIA, because insurers and future policyholders must pay all losses up to the market retention, so that lowering the deductible just transfers costs from the former to the latter with no effect on taxpayers.

As under TRIA, the cost to future policyholders is small relative to the cost to taxpayers for NBCR attacks (see Figure 4.4B). The hard cap also significantly limits the cost to insurers for very large NBCR attacks in those circumstances in which the existing cap is soft, with the result that taxpayers pay a very high proportion of losses in the largest attacks. What the figure does not show, however, is how payments by insurers, taxpayers, and future policyholders compare to those under TRIA for the smaller, likelier, NBCR attacks. We now address this issue.

Trade-Offs Between Taxpayers and the Insurance Industry

In Chapter Three, we found that expected annual taxpayer cost is lower with than without TRIA over a broad range of assumptions about the relative likelihood of large versus smaller attacks and the fraction of uninsured and unpaid insured losses paid by the government after an attack. Even though TRIA increases government payments for larger attacks, the higher take-up rate for property damage increases the insurance industry's role in paying losses in smaller and presumably more probable attacks, leading to a decline in taxpayer cost when attacks are weighted by their likelihood of occurring. We now investigate whether similar findings hold for TRIA with NBCR coverage. Again, we focus on the version of NBCR coverage with a hard cap and a 7.5 percent deductible, although we also present findings for other versions of the program.

Conventional Attacks

Even though the deductible is lower, the higher take-up rate in many scenarios causes taxpayer cost to fall relative to TRIA for most the conventional attack scenarios examined (recall from Figure 4.1 that the take-up rate for property coverage can rise to 100 percent). As shown in Figure 4.5, taxpayer cost is lower than under TRIA in 59 percent of the conventional-attack scenarios examined. As noted previously, lowering the deductible does not cause taxpayer cost to increase, because TRIA's market-retention requirement causes future policyholders to bear much of the cost of the reduced deductible. The decline in taxpayer cost is most pronounced for the smaller attacks. Due to the insurer deductible and the market retention, taxpayers pay little or nothing for such smaller attacks and often provide less postattack compensation, because a higher take-up rate reduces uninsured losses.

Recall that, compared with letting the program expire, TRIA increases the fraction of losses compensated by the insurance industry in 71 percent of the conventional-attack scenarios examined (see Figure 3.5 in Chapter Three). In contrast, relative to TRIA, requiring insurers to offer a policy that covers both conventional and NBCR attacks with a hard cap and a 7.5 percent deductible reduces the fraction of losses paid by the insurance industry in 90 percent of the conventional attacks examined (see

Figure 4.5
Taxpayer Cost for Conventional Attacks Under TRIA and TRIA with NBCR-Attack Coverage, Hard Cap, and 7.5 Percent Deductible

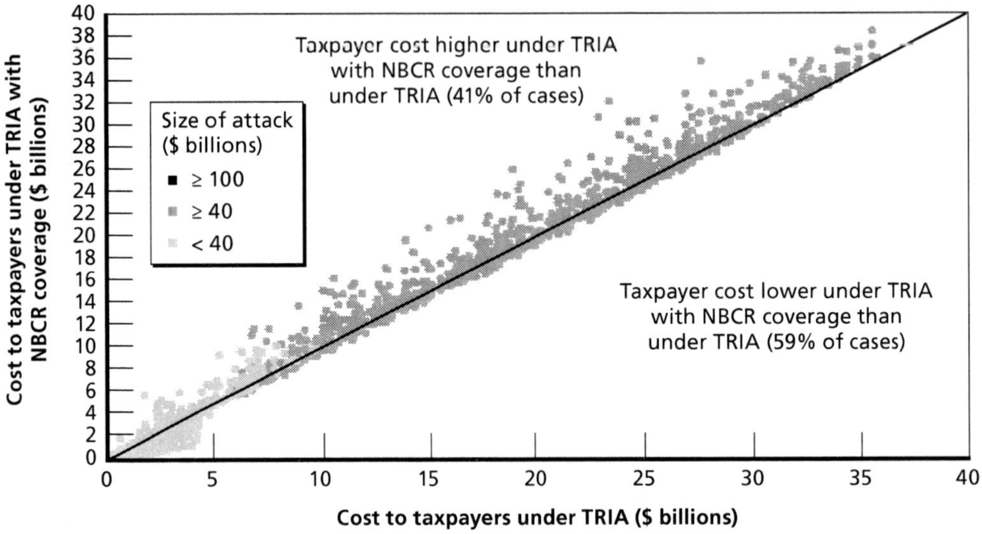

RAND *MG679-4.5*

Figure 4.6). The reduction of the insurer deductible from 20 percent to 7.5 percent is the primary reason for the diminished insurer role. The 10 percent of cases in which the insurer role is greater is due to cases in which the take-up rate is substantially above that observed with TRIA.

NBCR Attacks

The substantial increase in the take-up rate for NBCR attacks for TRIA with NBCR coverage compared with TRIA as currently configured reduces taxpayer cost in 33 percent of the NBCR scenarios examined (see Figure 4.7). Declines in taxpayer cost are restricted to the smaller NBCR attacks, attacks in which the TRIA cap does not come into play. Added government responsibility for insured losses between $100 billion and $650 billion means that taxpayer cost can rise substantially relative to TRIA for large NBCR attacks.

Because the NBCR take-up rate remains low under TRIA and, in many of the scenarios examined, TRIA's cap significantly limits insurer payouts for insured losses above the cap, TRIA either reduces or leaves unchanged the insurer role in paying losses due to NBCR events relative to letting the program expire (see Figure 3.7 in Chapter Three). Relative to TRIA, however, mandating the offer of a policy that covers both conventional and NBCR attacks results in the insurance industry playing a greater role in providing compensation in nearly one-third of the NBCR scenarios examined (see Figure 4.8). The effect is particularly pronounced for radiological bomb

Figure 4.6
Share of Losses Paid by Insurers for Conventional Attacks Under TRIA and TRIA with NBCR-Attack Coverage, Hard Cap, and 7.5 Percent Deductible

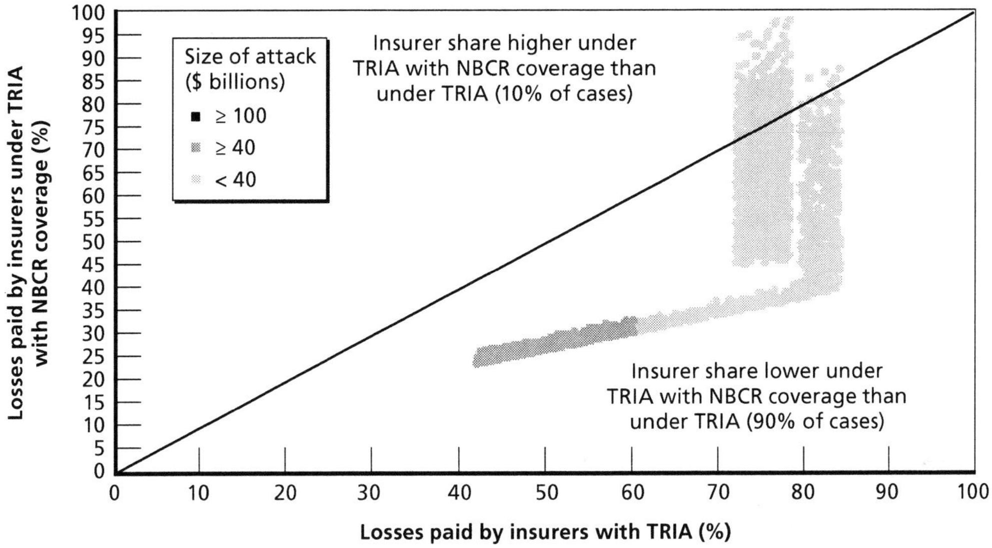

RAND *MG679-4.6*

Figure 4.7
Taxpayer Costs for NBCR Attacks Under TRIA and TRIA with NBCR-Attack Coverage, Hard Cap, and 7.5 Percent Deductible

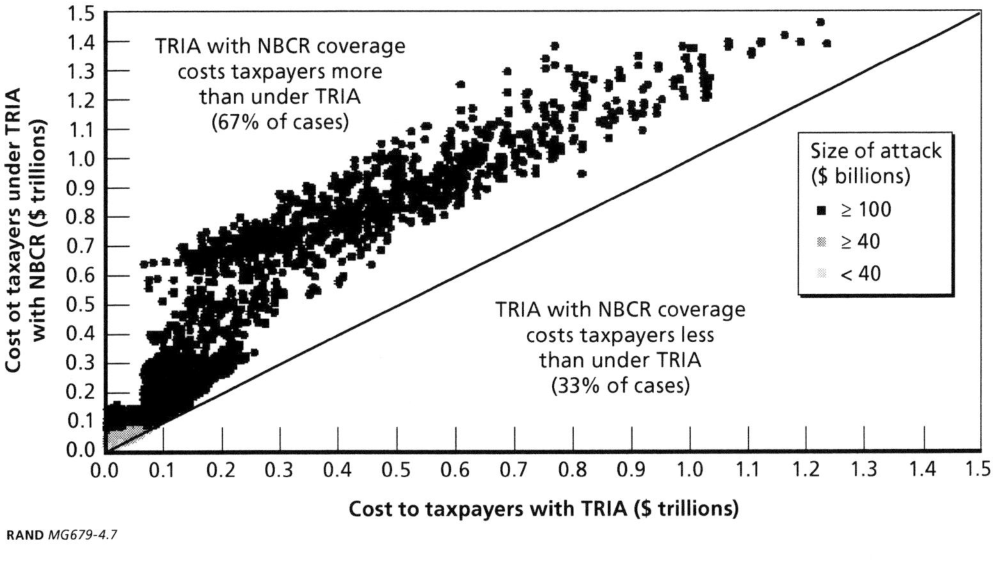

RAND *MG679-4.7*

attacks. These attacks are represented by the column of points above the 45-degree line in the lower left part of the chart. Radiological attacks produce almost entirely

Figure 4.8
Share of Losses Paid by Insurers for NBCR Attacks Under TRIA and TRIA with NBCR-Attack Coverage, Hard Cap, and 7.5 Percent Deductible

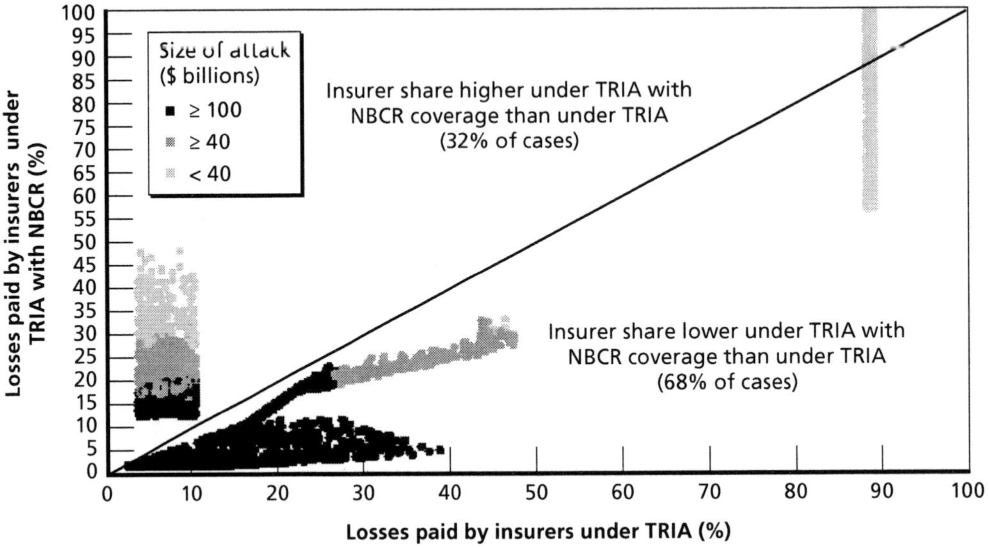

RAND *MG679-4.8*

property losses (as opposed to WC losses) and thus benefit the most from the higher take-up rates for NBCR attacks. Note, however, that the fraction of losses paid by the insurance industry is still not particularly high for any of the radiological attack scenarios simulated and remains near 10 percent for the largest radiological attacks.

The fraction of losses paid by the insurance industry declines in 68 percent of the NBCR scenarios examined, with some of the steepest declines for nuclear attacks (represented by the clump of points closest to the horizontal axis in Figure 4.8). Even though the take-up rate for nuclear attacks increases, reducing the deductible and hardening the cap reduce the amount that the insurance industry pays for these extremely large losses. The fraction of nuclear losses compensated by insurers under TRIA with NBCR coverage remains low, seldom exceeding 10 percent of total losses.[4]

Analogous to the fraction of losses paid by insurers with and without TRIA for conventional attacks, we find that TRIA with NBCR coverage tends to increase the fraction of losses paid by insurers relative to TRIA for smaller NBCR attacks and decrease it for larger attacks. The fraction of scenarios in which the share of losses paid by the industry increases is greater for NBCR attacks causing less than $40 billion in

---

[4] The vertical bar of points in the upper right quadrant of Figure 4.8 is the sarin attack scenarios. The proportion of losses paid by the insurance industry decreases for most of these points because (1) most of the losses are WC losses that were already covered under TRIA and (2) the insurer deductible has been reduced from 20 percent to 7.5 percent.

losses than it is for attacks causing more than $40 billion in losses. In addition, TRIA with NBCR coverage increases the industry share of losses relative to TRIA for all of those NBCR attacks in our analysis that caused less than about $6 billion in losses, as well as for some attacks causing more than this amount in losses. For all attacks causing less than $6 billion in losses, the substantial increase in the take-up of NBCR coverage on property policies combined with little or no surcharge on future policyholders or taxpayer payment through the government program cause the fraction of losses paid by insurers to increase relative to TRIA.

Table 4.1 summarizes the effect of all four versions of TRIA with NBCR coverage on taxpayer cost and the fraction of losses paid by insurers. Results for the version just discussed (hard cap and 7.5 percent deductible) are shown in the last row of the table. For NBCR attacks, the effects of the other three versions are more positive in terms of taxpayer cost and the fraction of losses paid by the industry (see the last two columns of Table 4.1). However, the improved performance for NBCR attacks is coupled with mixed or worse performance for conventional attacks in these dimensions and, sometimes, worse performance for NBCR or conventional attacks in other dimensions.

Expected Taxpayer Cost

The finding that taxpayer cost is lower under TRIA with NBCR coverage than it is under TRIA for the smaller attacks (Figures 4.5 and 4.7) raises the possibility that expected annual taxpayer cost will be lower than it is under TRIA when terrorist attacks are weighted by the probability of occurrence. However, considerable

Table 4.1
Comparison of Taxpayer Cost and Share of Losses Paid by Insurers Under TRIA and TRIA with NBCR-Attack Coverage

	Scenarios (%)			
	Conventional Attacks		NBCR Attacks	
Variant of TRIA with NBCR Coverage	Taxpayers Pay Less Under TRIA with NBCR Than Under TRIA	Industry Share Greater Under TRIA with NBCR Than Under TRIA	Taxpayers Pay Less Under TRIA with NBCR Than Under TRIA	Industry Share Greater Under TRIA with NBCR Than Under TRIA
Existing cap				
20% deductible	26	0	62	80
7.5% deductible	26	0	56	43
Hard cap				
20% deductible	42	37	36	64
7.5% deductible	56	10	33	32

NOTE: Perception of existing cap varies from hard to soft.

uncertainty remains about the probability of terrorism attacks, and, as in Chapter Three, we compare expected taxpayer cost over a wide range of assumptions about the probability of attack and the fraction of uninsured losses and insured but unpaid losses compensated by the government.

Figure 4.9 is analogous to Figure 3.8 in Chapter Three and distinguishes the conditions under which the expected taxpayer cost under TRIA with NBCR coverage, a hard cap, and 7.5 percent deductible is less than that under TRIA. If the relative probability of large and smaller attacks is the same as in the RMS model (a value of 1 on the horizontal axis) and the government compensates 25 percent or more of uninsured and insured but unpaid losses, expected taxpayer cost under TRIA with NBCR coverage, a hard cap, and a 7.5 percent deductible is less than under TRIA. The requirement on government compensation rises as the relative probability of large attacks increases, reaching approximately 65 percent if the relative probability of

Figure 4.9
Expected Taxpayer Cost for TRIA Compared with That for TRIA with NBCR-Attack Coverage, Hard Cap, and 7.5 Percent Deductible

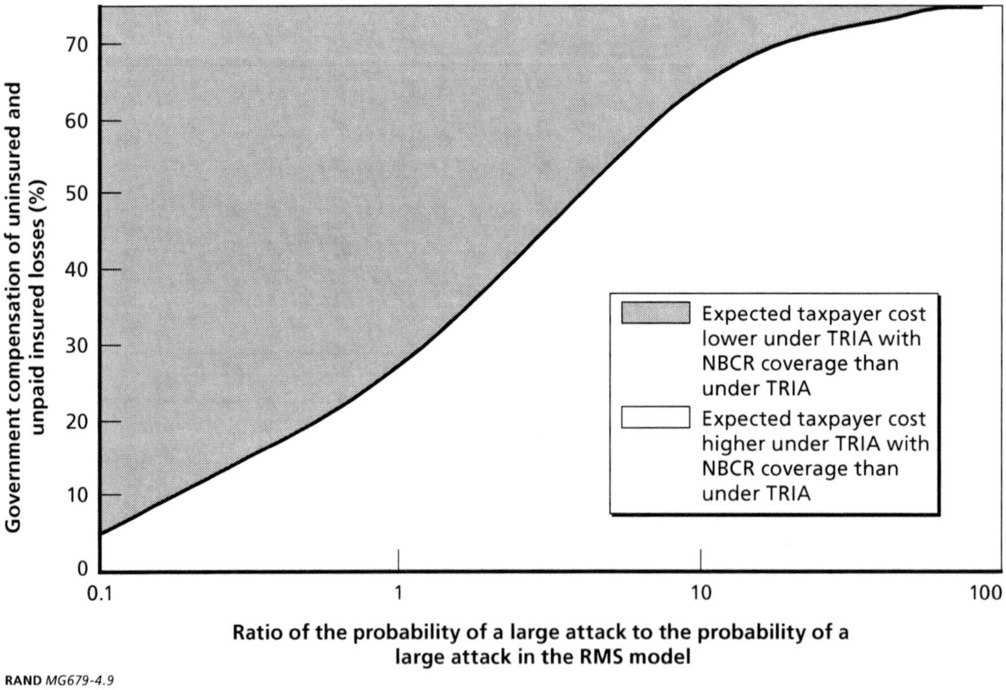

RAND *MG679-4.9*

large and smaller attacks is 10 times that observed in the RMS model.[5] Appendix E provides the relative probabilities of large and smaller NBCR attacks and the numerical results that underlie Figure 4.9.

We found in Chapter Three that expected taxpayer cost for NBCR attacks is equal to or higher with TRIA than without TRIA for all the scenarios involving NBCR attacks examined (see Figure 3.6 in Chapter Three). Thus, TRIA costs taxpayers more than does having no government program for NBCR attacks regardless of assumptions about the relative likelihood of large and smaller attacks or about government compensation for uninsured and insured but unpaid losses. Because the take-up rate for terrorism coverage on property policies for NBCR attacks increases under this version of TRIA with NBCR coverage, there are now combinations of relative attack probabilities and government compensation that result in lower expected taxpayer cost for NBCR attacks. Appendix E details such situations.

Extending TRIA to Cover NBCR-Attack Losses When the Cap Is Already Hard

So far, we have assumed that the cap of the current TRIA program varies from hard to very soft. We now examine the effect of requiring insurers to offer both conventional and NBCR coverage when insurers believe that the cap of the current TRIA program is hard, a special case of our more general assumption. If the cap is currently hard, insurers price coverage assuming that they have no responsibility to pay for losses once insured losses hit $100 billion. Thus, requiring insurers to offer NBCR coverage does not cause insurers' financial exposure to terrorism losses to increase nearly as much as it does when insurers believe that the TRIA cap is soft. Even though NCBR attacks can generate losses well in excess of $100 billion, the hard cap shields insurers from exposure to these very large losses.

Even though a preexisting hard cap limits exposure, extending TRIA to require offering both NBCR and conventional coverage without other changes in the program still does increase expected losses to insurers and thus causes premiums to increase and take-up to decrease. As shown in Figure 4.10, the take-up rate falls to between 30 and 60 percent if the deductible remains at 20 percent. The take-up rate for terrorism coverage on property policies can thus fall substantially from the 55 to 65 percent observed under TRIA, but the potential decline is not nearly as large as it is when the analysis includes scenarios in which the TRIA cap is soft (0 to 56 percent; see Figure 4.1).

---

[5] Expected taxpayer cost under TRIA with NBCR coverage is also lower than that under no government program under a broader set of conditions. For example, at the baseline attack probabilities, TRIA with NBCR coverage has a lower expected taxpayer cost if government compensation for uninsured losses and insured but unpaid claims exceeds approximately 15 percent versus the 25 percent in Figure 4.9. See Appendix E for more details.

Figure 4.10
Take-Up Rates for TRIA and TRIA with NBCR-Attack Coverage When the Current Cap Is Hard

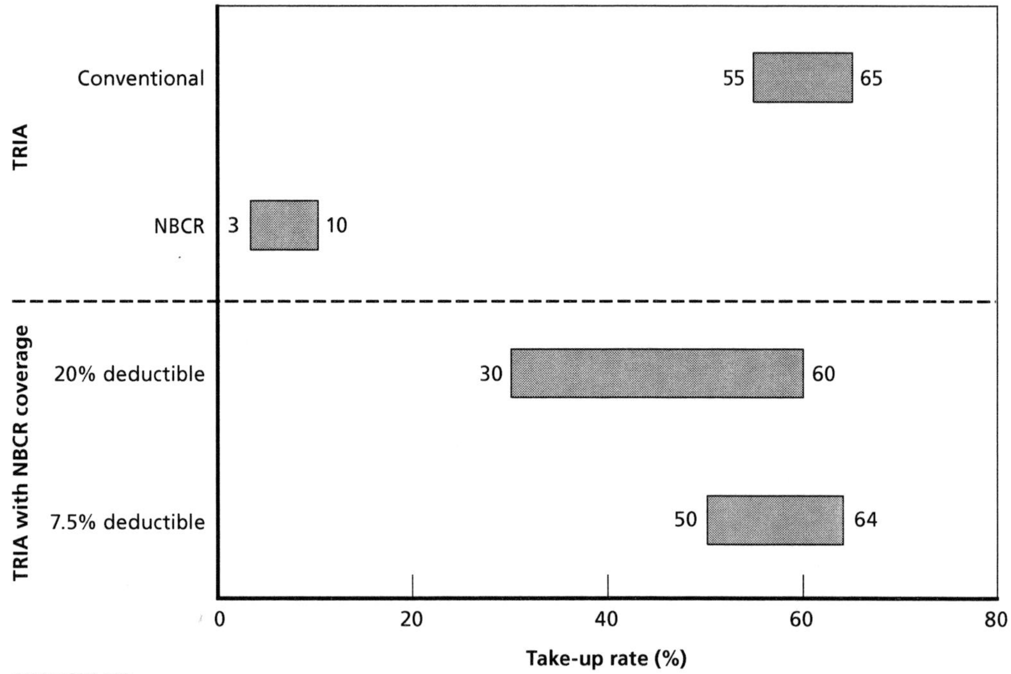

RAND *MG679-4.10*

If the deductible is lowered to 7.5 percent and the cap is already hard, the take-up rate ranges from 50 to 64 percent—not a large decline from the conventional take-up rate observed under TRIA. Recall from Figure 4.1 that both hardening the cap and lowering the deductible were necessary to produce take-up rates that did not decline much from the range under TRIA when the TRIA cap is perceived to be soft. This finding underlines the importance of lowering the deductible when extending TRIA to require an NBCR offer when the existing cap is hard or close to hard.[6]

Summary

Modifying TRIA to require insurers to offer policies that cover both NBCR and conventional attacks without changing other program features, such as the program deductible or the cap, has little upside for NBCR attacks and can have major, unintended consequences for conventional attacks. In particular, the outcomes for conven-

[6] Comparison of the penultimate intervals in Figures 4.1 and 4.10 implies that, if the existing cap is sufficiently soft, then hardening the cap is all that is required to prevent the take-up rate from falling below that observed under TRIA.

tional attacks when the cap and the deductible are unchanged are similar to those if TRIA expires. For NBCR attacks, this version of TRIA with NBCR coverage does not improve outcomes much over TRIA alone.

Hardening the cap and reducing the deductible to 7.5 percent generates outcomes comparable to or better than TRIA in several key dimensions for conventional attacks and improves outcomes in several dimensions for NBCR attacks (see Table 4.2). The take-up rate is higher in most of the conventional-attack scenarios examined, and the fraction of losses that remains uncompensated is lower. The cost to taxpayers falls in most of the smaller conventional-attack scenarios examined relative to TRIA but rises in most of the larger ones.

For NBCR attacks, the take-up rate rises substantially, and, as a result, the fraction of losses that remain uncompensated falls broadly across the NBCR-attack scenarios examined (see last two columns of Table 4.2). Even though the cap has hardened and the deductible is lower, the cost to taxpayers declines for most NBCR attacks, causing losses less than $40 billion, and the role played by the insurance industry in

Table 4.2
Summary of Outcomes for TRIA with NBCR-Attack Coverage, a Hard Cap, and a 7.5 Percent Deductible Relative to the 2007 Configuration of TRIA

	Conventional Attacks		NBCR Attacks	
Outcome Measure	**<$40 Billion**	**≥$40 Billion**	**<$40 Billion**	**≥$40 Billion**
Take-up rate	Most higher	Most higher	All higher	All higher[a]
Fraction of loss that remains uncompensated after an attack	Most lower	Most lower	All lower	All lower
Cost to taxpayers	Most lower	Most higher	Most lower	Most higher
Cost to future policyholders	Most higher	All higher	Most higher	All higher
Fraction of industry surplus used to compensate loss following an attack	Most lower	All lower	Most lower	Most lower
Fraction of loss paid by insurance industry	Most lower	All lower	Many higher[b]	Most lower

NOTE: Share of scenarios with indicated change in outcome. *Some* refers to 1 to 30 percent of scenarios. *Many* refers to 30 to 50 percent of scenarios. *Most* refers to 51 to 99 percent of scenarios.

[a] Excluding one very unusual scenario.

[b] Higher for all NBCR attacks causing losses of less than $6 billion.

compensating losses rises for all the NBCR attacks examined that caused less than $6 billion in losses. Future policyholders do fare worse in this intervention than they do under TRIA, because the lower insurer deductible leads to greater postattack surcharges.

As does the current version of TRIA for conventional attacks, TRIA with NBCR coverage, a hard cap, and lower deductible increases the industry role in providing compensation for smaller NBCR attacks in return for shifting some of the risk for larger attacks to the government. However, because the probability of large attacks is believed to be much lower than that of smaller attacks, expected taxpayer cost is lower for a program that hardens the cap, lowers the deductible, and requires NBCR coverage than it is under TRIA over a wide range of assumptions about the relative risks of large and smaller attacks and about the proportion of uninsured and unpaid insured losses compensated by the government. At current RMS estimates of large and smaller and conventional and NBCR attack probabilities, expected annual taxpayer costs are lower under TRIA with this version of NBCR coverage if the government compensates more than roughly one-quarter of uninsured and unpaid insured losses.

Due to the uncertainty over the hardness of the existing TRIA cap, our analysis suggests that both hardening the cap and lowering the deductible are critical to achieving positive outcomes when TRIA is expanded to require insurers to offer coverage for both NBCR and conventional attacks. If the existing cap is quite soft, lowering the deductible alone does not improve outcomes for NBCR attacks and can result in a deterioration of program performance for conventional attacks. If the cap is already fairly hard, hardening the cap would not make a great deal of difference, and lowering the deductible becomes key to avoiding adverse outcomes relative to TRIA. Given the measures considered in this monograph, hardening the cap and lowering the deductible together is thus robust to the substantial uncertainty over how insurers perceive the hardness of the current cap.

CHAPTER FIVE

Conclusion

Terrorism is an ongoing threat to life, property, and economic prosperity in the United States, and terrorism insurance is an important component of the system of institutions that can help mitigate the impact of the terrorism threat on U.S. society. A well-functioning terrorism insurance market can reduce the impact of terrorism on economic activity prior to any attack as well as enhance economic resiliency after any attack. However, debate exists over the role the federal government should play to ensure the proper functioning of this market.

This monograph contributes to this debate by comparing the performance of the current TRIA program and several alternative government interventions in the market for terrorism insurance over a wide range of scenarios that reflect important uncertainties about the insurance industry's preattack behavior, the current TRIA cap's hardness, the size and type of any terrorist attack, and any postattack government decisions to compensate uninsured losses.

Key Findings

Our analysis suggests important conclusions about the potential benefits of the current TRIA program as opposed to allowing the program to expire.

- TRIA has positive effects on the insurance market for conventional attacks: The proportion of property-insurance policies with terrorism coverage is higher and uncompensated losses are lower for conventional attacks with TRIA than they are without TRIA.
- For conventional attacks that cause less than about $40 billion in total losses, TRIA reduces taxpayer cost and increases the insurance industry's role in compensating losses relative to scenarios in which TRIA has expired. For attacks causing more than about $40 billion in damage, TRIA can significantly increase the cost to taxpayers and can reduce the proportion of losses paid by insurers relative to scenarios without TRIA. For comparison, note that the attack on the World Trade Center caused roughly $23 billion in insured property and WC losses.

- Because terrorism experts believe that larger attacks are much less likely than smaller ones, the expected cost to taxpayers is lower with TRIA than without TRIA over a wide range of assumptions about the likelihood of large attacks relative to smaller ones and about the proportion of uninsured loss that the government chooses to compensate after an attack.

The current TRIA program does not, however, provide similar benefits for NBCR attacks, which many see as a significant gap in the nation's ability to manage terrorism risk. We thus evaluated potential modifications to TRIA that require insurers to offer policies that cover losses due to both conventional and NBCR attacks. Our analysis concludes the following:

- Requiring terrorism policies to cover both conventional and NBCR attacks without changes in other program features may not improve outcomes much for NBCR attacks and may have significant adverse consequences for conventional attacks.
- Hardening the cap and lowering the deductible to 7.5 percent can improve outcomes for a program that requires insurer to offer both conventional and NBCR coverage. With such changes, the fraction of losses that remains uncompensated after NBCR attacks drops substantially from the level under TRIA.
- Due to uncertainty of the hardness of the existing TRIA cap, both hardening the cap and reducing the deductible are central to achieving positive outcomes for TRIA with NBCR coverage. If the existing cap is soft, hardening the cap is of critical importance. If the existing cap is hard, lowering the deductible is necessary. Given the outcome measures considered in this monograph, hardening the cap and lowering the deductible together is thus a robust strategy that effectively addresses the substantial uncertainty over how insurers perceive the hardness of the current cap.
- The expected cost to taxpayers may be lower for a program that requires NBCR coverage, hardens the cap, and lowers the deductible to 7.5 percent than taxpayer cost would be under TRIA if the government decides after an attack to compensate more than roughly a quarter of the uninsured loss.

Overall, TRIA and enhancing TRIA to require terrorism-insurance policies to cover both conventional and NBCR attacks achieve positive outcomes by transferring risk for the largest attacks to taxpayers. In return, the insurance industry can play a larger role in compensating losses for smaller attacks. Because the probability of large attacks is thought to be far lower than the probability of smaller attacks, both TRIA and TRIA with NBCR coverage can achieve these benefits while reducing expected taxpayer cost.

Implications of Findings for Recent Legislation

Our analysis provides some insights into legislation that the U.S. House of Representatives has recently passed to extend and modify the TRIA program (H.R. 2761). The bill requires insurers to offer coverage for conventional and NBCR attacks, includes detailed language that attempts to harden the program cap, and lowers the deductible for NBCR attacks, initially to 3.5 percent of an insurer's DEP. The deductible then gradually increases to 7.5 percent. (The deductible for conventional attacks remains at 20 percent.) While the interventions considered in this study differ in some important ways from this legislation, our analysis nonetheless provides some important insights.

The House bill includes several features identified in this study that will likely improve the performance of the TRIA program. First, the bill attempts to address the shortcomings of the TRIA program identified here for NBCR attacks. Second, the bill attempts to harden the TRIA cap, which our analysis suggests is important to successfully including NBCR coverage in the program. Finally, the bill lowers the deductible for NBCR attacks, consistent with the findings in our analysis.

Our analysis differs from the House bill in two important ways. First, the legislation attempts to harden the cap with detailed language and methods for prorating claim payments once insured losses exceed the cap, while the interventions considered here harden the cap by assuming that the government guarantees to pay insured losses from \$100 billion to \$650 billion. Critical to the House bill's impact on NBCR coverage will be insurers' perceptions about whether the bill's language is sufficiently strong to limit their actual liability for the insured losses over \$100 billion and how these perceptions evolve over time.

Second, the House bill links offers for NBCR and conventional terrorism coverage differently from how the options are linked in this study. As in our analysis, the House bill requires insurers to offer coverage for both conventional and NBCR attacks that does not differ in terms, amounts, or other conditions for coverage due to events other than terrorism. We require policyholders to either accept or reject this bundled coverage. Under the House bill, in contrast, if the policyholder rejects the initial offer of coverage, the insurer may offer coverage options that differ in terms, amounts, or other conditions from the underlying policy. In particular, an insurer may offer coverage for only conventional attacks and not NBCR attacks.[1] The question remains whether allowing policyholders to separately purchase conventional and NBCR coverage in the House bill will result in a sufficiently high take-up rate for NBCR coverage to generate outcomes like those found in this analysis. Existing research suggests that the demand

---

[1] That we modeled a different approach than the House bill should signal no policy preference. We settled on the options analyzed here before the House bill was introduced and chose to bundle NBCR and conventional coverage (1) because it seemed like the simplest way to extend TRIA to better address NBCR attacks and (2) because it is analytically more straightforward to analyze bundled coverage than it is to analyze unbundled coverage.

for NBCR coverage is low, and thus allowing this coverage to be offered separately may not result in substantial take-up. Further research on the effect of offering unbundled versus bundled conventional and NBCR coverage is clearly warranted. However, given the potential importance of this issue and the shortage of solid evidence on which to base any judgments, Congress should plan to review the effects of new legislation on NBCR take-up and revise its approach in the next few years as appropriate, even if it chooses to reauthorize the overall TRIA program for a longer period of time.

Moving Forward

This monograph has not addressed some issues relevant to a full assessment of government intervention in the market for terrorism insurance. For instance, we do not assess (1) the impact of changes in insurance price and take-up rate caused by TRIA and enhancements to it on economic activity preattack or the speed of economic recovery and resiliency of the economy after an attack, (2) how price changes might affect incentives for businesses to adopt measures to mitigate terrorism risk, (3) how any change in industry's existing willingness to bear terrorism risk might affect take-up rates over time, or (4) how government programs affect the flow of new capital into insurance markets or the development of instruments or strategies to spread insurance risk. Including such issues would increase the comprehensiveness of our analysis but would not affect the basic trade-offs identified here.

The models and methods used in this study are applicable to a wider range of questions than considered here. For instance, our simulations could be adapted to examine a broader range of modifications to the TRIA program, such as different insurer copayments or different program caps. These tools could also examine a broader range of government interventions in terrorism insurance markets, including requiring policyholders to purchase terrorism coverage and pooling arrangements in which a surcharge on insurance policies funds a pool that is then used to pay claims following any terrorist attack.

The threat of terrorism does not appear to be a transitory phenomenon confronting the United States, and the role that insurance can play in mitigating this threat warrants ongoing analysis.

APPENDIX A

RMS's Probabilistic Terrorism Model

We use RMS's Probabilistic Terrorism Model to provide the data for the terrorist attacks shown Table 2.4 in Chapter Two. The RMS model was developed primarily for use in the insurance industry to help property-casualty insurers manage their exposure to catastrophic terrorism loss. The model estimates the risk from a wide range of potential terrorist attacks and can be applied at any scale, ranging from an individual building to the entire country. The model computes the terrorism risk from the overall probability of an attack occurring, the relative likelihood of thousands of individual attacks, and the consequences of each attack.

The RMS model estimates the consequences, in terms of property damage and casualties, of various potential terrorist attacks based on the modeling of weapon effects and geocoded databases of structural characteristics of targets, population densities, human activity patterns, business activities, and the values of buildings and their contents. RMS estimates the overall probability of attack and the relative likelihoods of different types of attacks at different targets, using expert judgment about capabilities and objectives of terrorist groups; target selection by terrorists; capability requirements for different attack modes; and propensity to stage multiple, coordinated attacks. This appendix provides an overview of the RMS model. Additional information can be obtained from RMS (undated) or by contacting RMS directly.

Model Scope

The loss estimates in the RMS model focus on those losses that are normally eligible for insurance. Thus, the baseline estimates used in this study exclude losses that insurance does not normally cover but that may nonetheless be of interest to government policymakers, such as indirect economic losses (e.g., business losses resulting from decreased sales) and losses to entities that are typically self-insured (government property and employees). In addition, the losses modeled by RMS are restricted to those covered by commercial property and WC policies; the RMS model does not consider losses to noncommercial property (e.g., single-family homes and automobiles owned by individuals), nonemployee casualties (e.g., business patrons in stadiums, hotels, transportation systems, shops, or restaurants; visitors; passersby; people at home; people at school), psychological damage, latent injuries, or liability losses.

Estimating Attack Consequences

The RMS model estimates losses for a wide variety of different terrorist attack modes against a wide variety of different target types. The loss estimate in each of these terrorist attacks represents the total monetary loss from terrorist attacks, independent of the fraction of the total loss that insurance actually covers. Total terrorism losses estimated by the model include the sum of the replacement value of the property lost or damaged, the replacement value of the building contents lost, business interruption losses that would normally be eligible for insurance coverage, and WC insurance payments for employee death or injury. Property, contents, and business-interruption losses are based on values estimated on an individual-property and individual-business basis and are collectively referred to as *property losses* in our analysis. WC losses are based on the distribution of casualties across six severity categories and average WC insurance payments for each category in the state in which the attack occurs.

The consequences of each of these terrorist attacks are estimated from three components: weapon effects, target characteristics, and exposure characteristics. Weapon effects comprise the type of weapon, delivery mechanism, the hazards to people and property, and the spatial and temporal footprints of those hazards. A 600-pound bomb, for example, is detonated in a car, and damage occurs from blast pressure waves and debris impact that extends for tens of meters around the blast site.

The RMS model considers 37 different attack modes. Table A.1 provides a brief description of some general attack mode categories. For each attack mode in the model, RMS has developed physical event models that generate a hazard footprint that specifies a hazard level estimate as a function of location around the attack site. The size of the hazard footprint can vary from less than 100 meters (e.g., for a small bomb) to several hundred square kilometers (e.g., for a nuclear or outdoor biological attack).

The RMS model also includes 35 target types that are divided into eight groups representing distinct levels of threat. Table A.2 lists the target types included in each group.

Target characteristics include a number of building characteristics, such as height, number of stories, year built, and construction type, that influence the attack consequences. Characteristics may also include other factors specific to a particular target type. These target characteristics help define people's and structures' vulnerability to the hazard that the weapon imposes. For example, newer steel buildings will suffer less damage from a bomb than will older masonry buildings. Building

Table A.1
Modes of Attack Modeled in the RMS Terrorism Risk Model

Attack Mode Category	Description of Attacks in Model
Surface-to-air missile	Commercial 747 airliner shot down
Bomb	600 lb, 1 ton, 2 ton, 5 ton, and 10 ton
Aircraft impact	Hijacked 747 commercial airliner flown into a target
Conflagration	9,000-gallon gasoline tanker hijacked and set on fire
Sabotage: industrial, explosion	5-, 50-, and 150-ton TNT equivalents
Sabotage: industrial, toxic release	5%, 40%, and 100% of Bhopal accident
Sabotage: industrial, explosion + release	5 ton + 5% Bhopal, 50 ton + 40% Bhopal, and 150 ton + 100% Bhopal
Sabotage: nuclear plant, radiation release	0.5%, 5%, and 20% of inventory
Dirty bomb: cesium-137	1,500 Curies and 15,000 Curies
Chemical: sarin gas	Indoors: 10 kg; outdoors: 10 kg, 300 kg, and 1,000 kg
2% anthrax slurry released outdoors	1 kg, 10 kg, and 75 kg of slurry
Weaponized anthrax released indoors	40 g of weaponized anthrax
Smallpox	10, 100, and 1,000 initially infected
Genetically engineered smallpox	100 and 1,000 initially infected
Nuclear bomb	1 kiloton and 5 kiloton

Table A.2
RMS's Target Type Groups

Target Type Group	Target Types in Group
1	Government buildings
2	Business districts, skyscrapers, stock exchanges, hotels and casinos, airports, nuclear power plants
3	Military, train and subway stations, stadiums, bridges and tunnels
4	Industrial facilities, oil and gas, tourist attractions, shopping malls, restaurants, ports and ships
5	Media HQ, Fortune 100 HQ, theaters, major entertainment centers, gas stations
6	Cruise ships, apartment buildings, foreign consulates, United Nations
7	Water reservoirs and distribution, passenger trains, airspace zones
8	Power plants, dams, railway networks

SOURCE: RMS Probabilistic Terrorism Model.

characteristics are compiled from multiple sources, including data from the Sanborn Map Company.[1]

Attack exposure refers to the population and additional structures that an attack impacts. The exposure includes the number and spatial distribution of people within the hazard footprint, as well as (for insurance loss calculation purposes) their occupational status and age distribution. The exposure also accounts for the characteristics and density of structures within the hazard footprint. Along with those of the target itself, the characteristics of the exposure determine the losses from an attack in terms of the casualty distribution and property damage. Estimates of the number and demographics of building occupants are derived from local census data, journey-to-work data, building-use type, and building size. The number of occupants is also adjusted to account for the time of day. We examined the effects of midafternoon, weekday attacks. Most types of buildings would be most fully occupied at this time; therefore, our estimates reflect the worst case in the sense of the number of people exposed to the attack.

The model converts damage from an attack into losses in the form of casualties and property damage. The model provides estimates of the number of victims in each of six casualty categories: medical only; temporary, total disability; permanent, partial, minor disability; permanent, partial, major disability; permanent, total disability; and fatality. These categories correspond to the standard WC injury categories and are defined in the same way.

Property damage in the RMS model includes replacement value of damage to buildings and building contents and business-interruption losses. Building and building content losses represent the replacement value of damage to structures. Business-interruption losses represent losses resulting from a civil authority exclusion zone around the incident site; this includes losses only from business closure and does not include indirect losses such as decreased sales.

Estimating Attack Probability

In addition to estimates of the consequences of each attack, RMS also provides estimates of their likelihood. Given the deep uncertainty in these attack probabilities, we used them in this study primarily for two purposes. First, we used them to inform the take-up rate model's representation of industry's expectations about future losses.[2] Second, we used attack probabilities to inform judgments about the trade-offs among

---

[1] The Sanborn Map Company maintains spatial coordinates and numerous building attributes for buildings in more than 21 major metropolitan areas across the United States.

[2] Much of the industry does, in fact, use the RMS model to help inform decisions about what terrorism coverage to offer.

the three government interventions in terms of taxpayer cost. For instance, our analysis suggests that the cost to taxpayers under TRIA may be larger than the cost only if TRIA expires for attacks with total losses greater than approximately $40 billion. The RMS likelihood estimates suggest that the probability of such an attack is very small.

The RMS model bases its likelihood estimates on subjective judgments by experts (Morgan and Henrion, 1990). This is necessary because, fortunately, terrorist events have occurred infrequently compared to accidents and natural disasters. Given this sparse record, as well as the complex and dynamic social origins of terrorism, terrorism threat lacks a phenomenological basis from which to model attack probabilities quantitatively. This makes it difficult to use the historical patterns of terrorism to estimate future attack likelihoods.

The RMS model uses expert judgment to assess both the relative likelihoods of various attacks and the overall attack probability. Values for both the absolute attack probability and the relative likelihoods of individual attacks are derived through a structured expert elicitation process that is informed by terrorist attack histories and contextual trends such as mentions of particular cities and targets in extremist media. Expert elicitation conferences are held twice each year to ensure that probability profiles are consistent with the most current information and analysis. Details of how this process is carried out can be obtained from RMS.

Overall Probability of Attack

RMS develops the overall annual probability of a terrorist attack occurring from several components:

- the probability that a terrorist attack of any kind will occur in the next year
- the probability that, if an attack occurs, it will be a single attack or a set of coordinated attacks
- the probability that, if an attack occurs, there will be other attacks within the year
- the probability that an attempted attack will succeed.

RMS develops three probability estimates for different assumptions about the terrorist threat level. These estimates reflect differing interpretations of available intelligence and consider capabilities and objectives of terrorist groups, access to particular weapon types, and effectiveness of counterterrorism efforts. The three threat outlooks are summarized next (RMS, 2003)

Reduced Threat Outlook. Optimistic interpretations of the available intelligence that imply a low risk of terrorist loss in the United States. Al Qaeda attack probability is assumed to be reduced from its long-term historical average. Less destructive attack modes are likelier, and the chance of an al Qaeda NBCR attack is negligible. Other foreign threat groups will not be active.

Standard Threat Outlook. Best assessment of the risk of large-scale terrorism loss in the United States throughout the year, resulting from all known terrorism threat groups. Probability of attack from al Qaeda is below its long-term historical worldwide average. Medium-scale attack modes predominate, and the chance of a NBCR attack is small. There is the additional possibility of attacks from other foreign threat groups.

Increased Threat Outlook. Pessimistic interpretations of the available intelligence that imply a heightened risk from terrorist loss in the United States during the year. Al Qaeda attack probability is assumed to be similar to its long-term historical average. Destructive attack modes are likely, and the chance of an al Qaeda NBCR attack is significant. There is the additional possibility of attacks from other foreign threat groups.

Relative Likelihoods of Attacks

Several factors influence the relative likelihoods of terrorist attacks in the RMS model. These factors can be classified into four components:

- the relative likelihood that any particular city will be attacked
- the relative likelihood that any particular target type will be attacked
- the relative likelihood that any specific target will be attacked because of its inherent iconic value or security
- the relative likelihood that any particular attack mode will be used in an attack.

Relative likelihoods define the probability that, if an attack happens, it will be in a particular place, of a particular type, on a particular target type, and, in some cases, on a particular target.

City Tier Likelihood. The RMS model groups cities into eight tiers according to relative likelihood of attack. The terrorist attack risk in the RMS model is heavily concentrated in a small number of cities, illustrated by the fact that the likelihood of attack in a city not ranked in the top 10 is estimated to be less than 11 percent.

Target Type and Individual Target Likelihoods. As with cities, RMS bins target types into separate groups according to their relative likelihood of attack (see Table A.2). The RMS model also provides the ability to incorporate attack likelihoods for specific individual targets based on their iconic value and security status. The iconic value parameter allows the attack likelihood for individual high-profile targets, such as iconic buildings, to be increased. Conversely, the security parameter allows the attack likelihood for individual targets with particularly high security, such as the White House, to be decreased. In general, however, this feature is largely unutilized in the RMS model because the specific information needed to assess these parameters for individual buildings is not available. That is, nearly all targets of a given target type are assigned the same iconic value and security levels.

Attack Mode Likelihood. The RMS model also assigns each attack mode, such as those shown in Table A.1, a relative likelihood based, in part, by the notion of each mode's "logistic burden." The logistic burden is a cost assigned to each mode that reflects skill, labor, time, and financial resource requirements. More resource-intensive modes have a higher logistics burden, which decreases their relative likelihood.

The relative likelihood of a terrorist attack depends strongly on the type of attack being considered. The variation in relative likelihood by attack mode spans several orders of magnitude. This large range in attack mode likelihoods is broadly consistent with the variation in attack mode likelihoods seen in the historical terrorism record (LaTourrette et al., 2007).

Annual Loss Exceedance Probability Distributions

Uncertainties in the individual parameter values used in the model give rise to an overall uncertainty in the terrorism risk estimates. This uncertainty is characterized in terms of a probability distribution of the annual loss exceeding any value. The exceedance probability distribution accounts for uncertainties in both attack likelihoods and attack consequences, which include variations in the threat outlook, hazard distribution (e.g., blast pressure transmission), vulnerability (e.g., the extent of building damage), as well as uncertainties in model parameters and data.

The annual exceedance probability distributions for conventional and NBCR attacks are shown in Figure A.1. The expected value for annual loss is the average value along the exceedance probability distribution and is $1.4 billion for conventional attacks and $0.32 billion for NBCR attacks.

Figure A.1
Exceedance Probability Distributions for Conventional and NBCR-Attack Losses

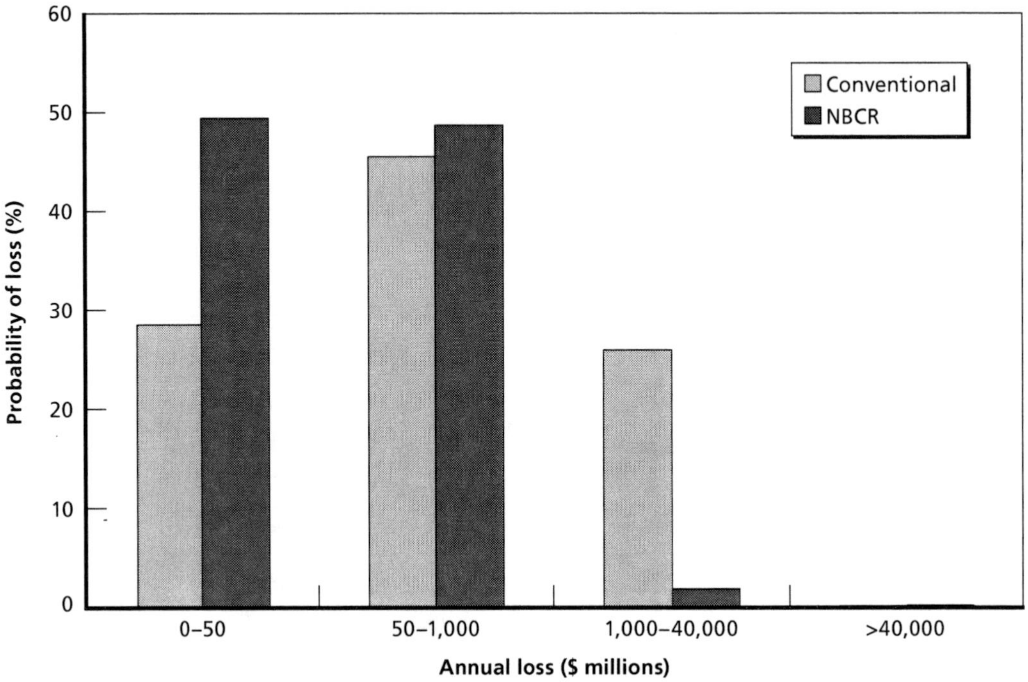

RAND *MG679-A.1*

APPENDIX B

Policyholder Take-Up of Terrorism Insurance

Introduction

This appendix describes the approach used to predict how take-up rates for terrorism insurance will change if TRIA were to expire and if TRIA were enhanced by requiring insurers to offer coverage for NBCR attacks and conventional attacks. First, a model of the cost of writing terrorism coverage is developed. The cost model is developed in several steps. It starts with a probability distribution for the total commercial property and WC losses (insured and uninsured) caused by terrorist attacks (see Appendix A). A baseline estimate of the take-up rate for terrorism coverage is then used to determine insured losses industrywide, and assumptions are made about the claims that an individual insurer would expect. The cost of writing terrorism coverage is subsequently based on the insurer's expected claim payments and the cost of the capital needed to cover claims should the costs exceed their expected levels. The next section describes how this cost model is used to project changes in take-up rates from changes in policy regime. Changes in the cost of writing terrorism insurance coverage when the policy regime changes (for example, from TRIA to no government program) are translated into changes in the price of insurance, which then drives the change in take-up rate for terrorism coverage on property policies.

Insurer Cost Model

Total Losses Caused by Attacks

The RMS model was used to develop a probability distribution of the losses from foreign-sponsored acts of terrorism in the United States.[1] The RMS model produces an exceedance probability curve of the form

[1] The vast majority of the attacks in the RMS model are in high-risk areas. Thus, we interpret the output of the RMS model as characterizing terrorism risk in high-risk areas.

$$\text{Prob}\left(L > L_j^e\right) = p_j^e,$$

where

L is the total WC and property losses at commercial locations, regardless of whether there is terrorism coverage for property losses ($ billions),

L_j^e is loss j reported in the RMS exceedance probability analysis,

j is the index of the losses reported in the RMS exceedance probability analysis, $j = 1, 2, \ldots 5{,}100$, and

p_j^e is the probability that the loss will exceed L_j^e.

We took the difference between successive values of p_j^e and averaged successive values of L_j^e to create a discrete probability distribution (P_J, L_J), which specifies the probability of an event of size L_J occurring for $j = 1, 2, \ldots 5{,}099$

Insured plus uninsured losses were then decomposed into WC and property losses based on analysis of the different types of terrorist attacks represented in the RMS model.[2] The fraction of loss L that is from NBCR attacks was found to be a function of the size of the loss:

$$\eta_{NBCR} = \begin{cases} 0.09 & L < 12.5 \\ 0.0164L - 0.115 & 12.5 \le L < 37.5 \\ 0.00192L + 0.428 & 37.5 \le L < 100 \\ 0.62 & L \ge 100 \end{cases}.$$

Similarly, the fraction of loss that is a property loss (as opposed to a WC loss) was also found to be related to the size of the loss:

$$\eta_{Prop} = \begin{cases} 0.8 & L < 1.0 \\ 0.75 & 1.0 \le L < 10. \\ 0.55 & L \ge 10 \end{cases}$$

Based on these relationships, the overall loss thus breaks down as follows:

[2] The various attacks RMS models were sorted into bins by size based on overall loss (insured plus uninsured). The proportion of losses in each bin (1) from NBCR attacks and (2) that were property losses was then calculated. Linear interpolations were constructed between the values determined for each bin.

$$L_{WC} = \left(1 - \eta_{Prop}\right) L$$
$$L_{P,Conv} = \eta_{Prop}\left(1 - \eta_{NBCR}\right) L$$
$$L_{P,NBCR} = \eta_{Prop}\eta_{NBCR} L,$$

where

L_{WC} is WC loss when total loss is L,

$L_{P,Conv}$ is property losses from conventional attacks when total loss is L, and

$L_{P,NBCR}$ is property losses from NBCR attacks when total loss is L.

WC losses are not decomposed by type of attack, because WC policies cover loss regardless of cause.

Industrywide Insured Losses Generated by Attacks

Assumptions about take-up rates for terrorism insurance are used to determine the insured property losses that will result from the total losses specified previously. In certain states, property losses are covered that result from fire, regardless of the underlying cause of the fire (as long as there is a property policy in place). Thus, even if a property policy excludes losses from terrorist attacks, property losses from a fire caused by an attack would be covered. Based on analysis of the losses from a fire for the attacks represented in the RMS model, we developed the following relationships for the fraction of property losses from fire following for a terrorist event of size L:

$$f_{Conv} = 0.094 \text{ for all } L$$
$$f_{NBCR} = \begin{cases} 0.0002 & \eta_{NBCR} L < 10 \\ 0.0025 & 10 \le \eta_{NBCR} L < 25 \\ 0.0096 & 25 \le \eta_{NBCR} L < 100 \\ 0.023 & \eta_{NBCR} L \ge 100 \end{cases}.$$

The take-up rate for terrorism insurance and the fire-following factors are used to project the insured losses for the insurance industry as a whole.

$$I_{WC} = L_{WC}$$
$$I_{P,Conv} = \left(\gamma_{Conv} + f_{Conv}\left(1 - \gamma_{Conv}\right)\right) L_{P,Conv}$$
$$I_{P,NBCR} = \left(\gamma_{NBCR} + f_{NBCR}\left(1 - \gamma_{NBCR}\right)\right) L_{P,NBCR},$$

where

I_{WC} is WC losses for loss of size L,

$I_{P,Conv}$ is insured property losses from conventional attacks for loss of size L,
$I_{P,NBCR}$ is insured property losses from NBCR attacks for loss of size L,
γ_{Conv} is the terrorism-insurance take-up rate for property policies when losses are from conventional attacks, and
γ_{NBCR} is the terrorism-insurance take-up rate for property policies when losses are from NBCR attacks.

The above equations reflect the fact that WC insurance policies cover losses regardless of the cause and our assumption that property losses not covered by a terrorism policy are covered in part from the fire-following regulations.

Total insured loss produced by an attack L is then $I = I_{WC} + I_{P,Conv} + I_{P,NBCR}$.

An Individual Insurer's Share of Industrywide Insured Losses

Not all insurers in the industry will have policyholders that are affected by a terrorist attack. Assumptions about the percentage of insurers affected by an attack and the distribution of insured losses across those affected are used to project the insured losses expected by an individual insurer for an attack of size L. Insurance Services Office (ISO), a leading firm that gathers data and analyzes claims and prices in the insurance industry, developed estimates of the percentage of the premium base that would be affected by terrorist attacks of different sizes (ISO, 2002, p. 20). We interpolated between these estimates to derive

$$\beta^I = \begin{cases} 0.1I + 0.05 & I < 1 \\ 0.05I + 0.1 & 1 \leq I < 10 \\ 0.00111I + 0.589 & 10 \leq I < 100 \\ 0.7 & I \geq 100 \end{cases},$$

where β^I is the proportion of industry premiums accounted for by insurers with policyholders affected by an attack of size L.

An individual insurer is assumed to have pessimistic expectations about the incidence of claims across insurers. In particular, we assume that an insurer expects that its policies will be hit first and that other insurers will be brought in only as the losses grow sufficiently large.

$$\beta^i = \begin{cases} 1 & \beta^i < \beta^i_m \\ \dfrac{\beta^i_m}{\beta^I} & \beta^I \geq \beta^i_m \end{cases},$$

where

β^i is the share of total insured losses borne by insurer i, and

β^i_m is the market share of insurer i as measured by its share of total industry premiums in TRIA lines.

Interviews with insurers conducted during the course of this study motivate our decision to assume that individual insurers have pessimistic expectations about the incidence of claims across insurers. Specifically, when modeling its exposure to terrorism losses in a particular city, an insurer typically searches for the attack location that would cause the largest insured losses to its policyholders. Such behavior is similar to assuming that its policyholders will be affected first by an attack.

Insurer Claim Payments for an Attack of a Given Size

The overall amount of insured losses generated by an attack, the share of insured losses borne by the individual insurer, and the characteristics of the government program for terrorism insurance determine an insurer's expected claim payments.

$$P^i = \begin{cases} \beta^i I & \beta^i I < D^i \text{ and } I < K \\ D^i + c_1\left(\beta^i I - D^i\right) & \beta^i I \geq D^i \text{ and } I < K, \\ P^i_K + c^e_2\left(\beta^i I - P^i_K\right) & I \geq K \end{cases}$$

where

P^i is a claim payment by insurer i for a loss of size L,

D^i is the deductible of insurer i under the government terrorism-insurance program

K is the cap on annual liability for insured losses established by the government terrorism-insurance program,

c_1 is the insurer copayment for insured losses above its deductible (D^i) until total insured losses industrywide reach K,

c^e_2 is the expected insurer copayment for insured losses when insured losses are above K,

P^i_K is the amount that insurer i pays when industry-insured losses equal K,

β^i_K is the percentage of insured losses borne by insurer i when industry-insured losses equal K, and

$$P^i_K = \begin{cases} \beta^i_K K & \text{for } \beta^i_K K < D^i \\ D^i + c_1(\beta^i_K I_K - D^i) & \text{for } \beta^i_K K \geq D^i \end{cases}.$$

The ratio of P^i to the total insured losses faced by the insurer, $\beta^i I$, represents the fraction of insured losses that the insurer will pay, given the government terrorism insurance program. We assume that both WC claims and property claims against the insurer are reduced by this ratio. Thus,

$$P^i_{wc} = \frac{P^i}{\beta^i I}\beta^i I_{WC} = \frac{P^i}{I} I_{WC}$$

$$P^i_p = \left(\frac{P^I}{\beta^i I}\right)\beta^i \left(I_{P,NBCR} + I_{P,Conv}\right) = \left(\frac{P^i}{I}\right)\left(I_{P,NBCR} + I_{P,Conv}\right),$$

where

P^i_{wc} is payments on WC claims by insurer i for a loss of size L, and

P^i_p is payments on property claims by insurer i for a loss of size L.

Cost of Writing Terrorism Coverage

An insurer's cost of writing terrorism coverage includes the expected payments multiplied by an expense factor that reflects the costs of writing policies and adjusting claims. The cost also includes the cost of the capital needed to ensure, with a given probability, that the insurer has enough resources to finance claim payments should they exceed expected payments.[3] The insurer selects this planning probability, with a lower probability implying that the insurer must obtain more capital:

$$C^i_{wc}(D^i, c_1, c^e_2, K, p^e) = hE^i_{wc} + \lambda(P^i_{wc,p^e} - E^i_{wc})$$

$$C^i_p(D^i, c_1, c^e_2, K, p^e, \gamma_{Conv}, \gamma_{NBCR}) = hE^i_p + \lambda(P^i_{p,p^e} - E^i_p),$$

where

C^i_{wc} is the cost to insurer i to support terrorism coverage on its WC policies,

C^i_p is the cost to insurer i to support terrorism coverage on its property policies, given the specified take-up rates for terrorism coverage on property policies for conventional and NBCR attacks,

h is the expense factor ($h \geq 1$),

λ is the cost of capital ($\lambda < 0$),

P^i_{wc,p^e} is WC payments by insurer i when the exceedance probability is p^e,

P^i_{p,p^e} is property claim payments by insurer i when the exceedance probability is p^e,

E^i_{wc} is expected WC payments by firm i for losses due to terrorist attacks,

E^i_p is expected property payments by firm i for losses due to terrorist attacks,

---

[3] For literature motivating this approach, see Harrington and Niehaus (2004, pp. 82–89, 146–147), who discuss the value of holding capital to reduce the probability of insolvency from claim costs exceeding insurer assets; Dong, Shah, and Wong (1996, p. 213), who present capacity-based pricing formulas that are a linear function of expected loss and the standard deviation or variance of losses; and Cummins (1991, pp. 267–272), who discusses situations in which an insurer will charge expected loss plus a premium to establish a buffer fund needed to avoid the risk of ruin.

$$E^i_{wc} = \sum_{j=1}^{5,099} p_j P^i_{wc,j}, \text{ and}$$

$$E^i_p = \sum_{j=1}^{5,099} p_j P^i_{p,j}.$$

Note that the amount of government payments under TRIA that is recouped from policyholders does not enter the insurer cost function. Policyholder surcharges (surcharges on commercial property-casualty policyholders, not just those that purchase terrorism coverage) finance such reimbursements, and we ignore the effect of such surcharges on the demand for insurance.[4] Thus, government decisions about what portion of its outlays to recoup from policyholders do not affect insurer costs for terrorism coverage in this model.

Note also that we allocated the capital required to avoid bankruptcy to the terrorism insurance line. Capital may be able to be shared across lines, and previous research has shown that the line-by-line allocation of capital should be determined with reference to the overall portfolio of lines (see, for example, Myers and Read, 2001). We have ignored such diversification opportunities in our model. However, when we calibrate the model, the planning probabilities are, in effect, adjusted to account for the actual sharing of capital that occurs across lines.

The fire-following regulations in some states mean that insurers will be subject to some terrorism risk on property policies, whether or not policyholders purchase terrorism coverage. (WC policies, in contrast, always cover terrorism losses.) Consequently, some costs of terrorism coverage should be attributed to property policies regardless of whether additional terrorism coverage has been purchased, and other costs should be attributed to the costs of additionally purchased terrorism coverage. To determine the cost attributable to the separate terrorism coverage, we first determine the cost of terrorism risk for property policies assuming that the take-up rate for terrorism coverage is zero. Then, we calculate the cost when the take-up rate for explicit terrorism coverage is set at the value assumed for the insurer. The difference between the two costs is the cost attributable to the terrorism policies.

Effect of the Change in Government Program on Take-Up Rate

We now use the cost model to estimate the impact of a change in government intervention (with the 2007 version of TRIA as the base case) on the take-up rate. We first describe the method for calculating the change in take-up rate from the change in

[4] We are also ignoring any management costs that insurers incur in collecting the surcharge that might be passed on to policyholders through higher rates.

costs. We then detail the parameter ranges used to construct the scenarios for the take-up rate analysis, followed by the methods used to parameterize the three government interventions. We move on to describe how the existing literature on take-up rates with and without TRIA is used to calibrate the model and, finally, how the model is used to project take-up rates if TRIA is expanded to cover NBCR terrorism risk.

Calculating Change in Take-Up Rate from Change in Costs

Insurers are assumed to set the premium required equal to the cost projected from the cost model, and an elasticity of demand is used to translate the percentage change in cost for property policies to the percentage change in the take-up rate on property policies.[5] In the case of allowing TRIA to expire, the change in take-up rate is derived using the following:

$$\frac{\gamma_{conv}^{NoProg} - \gamma_{conv}^{TRIA}}{\gamma_{conv}^{TRIA}} = \varepsilon_d \frac{C_p^{NoProg} - C_p^{TRIA}}{C_p^{TRIA}},$$

where

C_p^{NoProg} is the premium that insurers require when parameter values are set to reflect the absence of any government intervention in the terrorism insurance market,

C_p^{TRIA} is the premium that insurers require for property policies when parameter values are set to reflect the 2007 TRIA program, and

ε_d is the elasticity of demand for the inclusion of terrorism coverage on property policies.

The change in the cost of providing terrorism insurance from a change in government intervention (and, consequently, the implied change in take-up rate) is calculated at the take-up rate observed under TRIA. Figure B.1 helps to clarify the implications of this approach in the case of allowing TRIA to expire. The demand for terrorism insurance on property policies from insurer *i* is represented by the downward-sloping demand curve. The line S_{TRIA} represents the supply curve under TRIA. The firm supply curve in this model would be flat (average cost would be constant) were it not for the TRIA deductible and the industry cap. Simulation of the cost model over a range of parameter values suggests that the average cost curve slopes slightly downward. By fixing the take-up rate at the values observed for TRIA in calculating the change in cost, we are estimating the change in price (and the implied change in take-up), using the distance AB in the figure. If the supply curve is downward sloping, the change in

[5] Setting the premium required equal to the cost of providing the coverage is equivalent to average cost pricing. The costs used to calculate the percentage change in costs are those that represent the cost of explicit terrorism coverage. Changes in costs that are spread across all property policies from fire-following regulations have been netted out.

Figure B.1
Approach Used to Calculate Change in Take-Up Rate

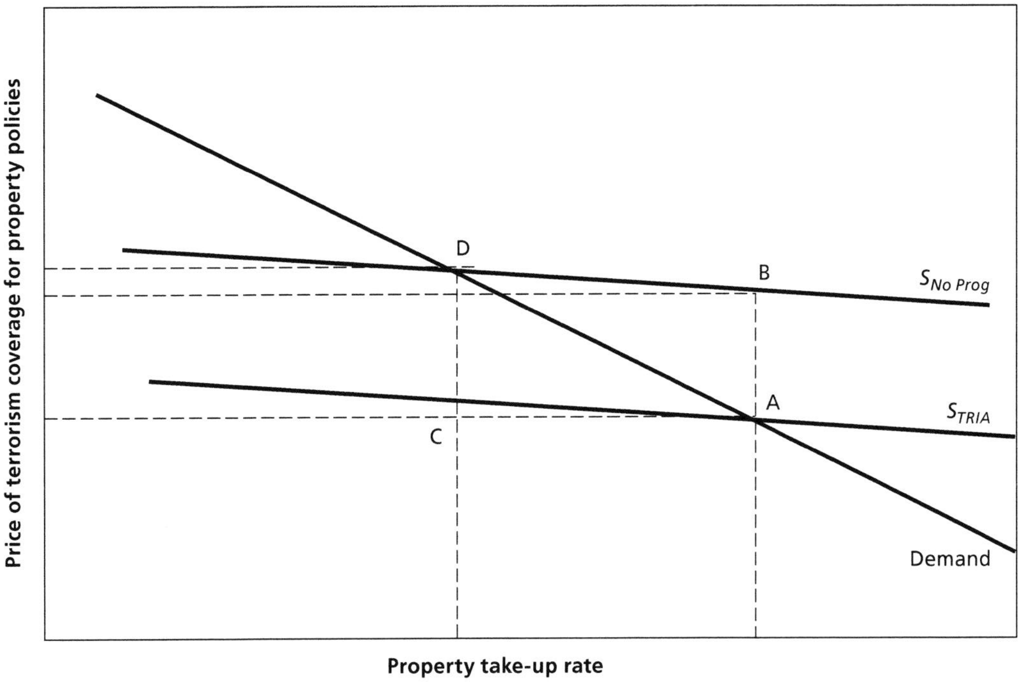

RAND *MG679-B.1*

price would be larger (CD) according to this model, and, thus, the change in take-up rate would be larger than predicted. Our analysis, however, suggests that, because the supply curve slopes downward only gradually, CD is not likely to be much larger than AB and that our approach provides a good first-order approximation of the change in take-up rate from a cost increase from C to D.

Parameter Ranges Used to Construct Scenarios for Take-Up Rate Analysis

The change in take-up rate is calculated for a large number of plausible futures. Each of these futures is characterized by a particular set of values for the parameter listed in Table B.1. The ranges considered for each of the parameters are included in the table (also reproduced in Table 2.3 in Chapter Two) and discussed below. The methods for selecting the scenarios are discussed in Chapter Two.

Take-Up Rate on Property Policies Under TRIA for Conventional Attacks. The range for the conventional take-up rate for property policies, 55 percent to 65 percent, is based on a variety of estimates in the literature. Insurance broker Marsh reported a 58 percent take-up rate for its clients (which are predominantly large companies) for policy renewals and purchases in 2005 and a 64 percent take-up rate in the fourth quarter of 2005 (Marsh, 2006, p. 7). Aon, an insurance broker that also serves large clients, put the take-up rate at 59 percent during the first two quarters of 2005 (Aon,

Table B.1
Parameter Values Varied in Experimental Design to Determine Change in Take-Up Rate

Parameter	Symbol	Possible Values	Basis for Values
Property take-up rate under TRIA for conventional attacks (%)	γ_{Conv}	[55, 65]	Surveys by U.S. Department of the Treasury and public reports by insurance brokers
Property take-up rate under TRIA for NBCR attacks (%)	γ_{NBCR}	[3, 10]	Surveys by U.S. Department of the Treasury and the Risk Insurance Management Society
Exceedance probability used in insurer planning	p^e	1%, 0.1%, 0.01%, $500 billion event	Wide range of plausible values
Elasticity of demand	ε_d	[–0.975, –0.325]	Study by the Wharton School
Expense factor	h	[1.43, 2.0]	Typical expense factors observed in insurance
Cost of capital ($)	λ	[0.05, 0.25]	Range of plausible values
Insurer copayment above program cap (%)	c_2^e	[0, 75]	Range of plausible values
Market share of firm (%)	β_m^i	5	Not allowed to vary; varying value makes little difference

2005, p. 26). Based on a nationwide survey of policyholders, the U.S. Department of the Treasury (2005, p. 84) found that 58 percent of policyholders purchased some amount of terrorism coverage in 2004.

Our analysis of the effects of TRIA expiring takes the 2007 version of TRIA as the base case, but the take-up rates discussed here apply to 2005 and 2003. The insurer deductible and copayment increased between 2005 and 2007, tending to push terrorism costs up. However, the insurer deductible declined between 2005 and 2007 because the 2005 TRIA extension narrowed the insured lines that TRIA covered.[6] We used the cost model to analyze the effect of these changes on insurer costs and found that the opposing factors roughly cancelled each other out.[7] Thus, it seems reasonable to set the take-up rates in 2007 to those observed in 2005.

6 Commercial automobile insurance, burglary and theft insurance, surety insurance, professional liability insurance, and farmowners' multiple peril insurance were dropped from TRIA by the Terrorism Risk Insurance Extension Act (PWG, 2006, p. 12).

7 Specifically, we found that the cost of providing terrorism coverage, as predicted by the cost model, was very similar for the 2005 and 2007 versions of TRIA.

Take-Up Rate Under TRIA for Nuclear, Biological, Chemical, and Radiological Attacks. The range of NBCR take-up rates, 3 percent to 10 percent, is also based on empirical data. Based on policyholder surveys, the U.S. Department of the Treasury (2005, p. 106) found that less than 3 percent of policyholders reported NBCR coverage in lines other than WC in 2003, 2004, and 2005. A survey of corporate risk managers in 2006 conducted by the Risk and Insurance Management Society found that less than 10 percent of companies had NBCR coverage on property policies (PWG, 2006, p. 78).

The U.S. Department of the Treasury (2005, p. 88) includes insurance coverage provided by captive insurers and, thus, should, in principle, capture NBCR coverage from all insurance sources.[8] Growth in the number of captives since 2005, however, may mean that 2005 estimates of NBCR take-up may understate take-up in 2007. Aon (2006, p. 24) notes that access to NBCR coverage is *the* driving force behind the use of onshore captive insurers, and the U.S. Department of the Treasury (2005, p. 88) found that the use of captives for terrorism-risk insurance rose from about 3 percent of all policyholders in 2004 to almost 8 percent in 2005. We are not aware of data on how the use of captives has grown since 2005 and, thus, have used the 2005 estimates of NBCR take-up in our analysis.

Exceedance Probability. We selected a range of values that reflects a wide range of insurance-firm behavior. An exceedance probability of 1 percent (1 in 100 chance) usually results in little or no predicted change in take-up when TRIA expires, so there was not much point in selecting a higher value. Exceedance probabilities lower by one and two orders of magnitude are also considered. In addition, we project terrorism policy costs when insurers are planning for a $500 billion event, because some of the insurers we interviewed during the study said that they were concerned about a $500 billion event. There is, of course, great uncertainty in the exceedance probabilities out in the tail of the loss distribution. Assigning more weight to the very large losses in the RMS model would be equivalent to increasing the exceedance probabilities used in this analysis.

Demand Elasticity. Based on data from Aon, a Wharton School study estimates that the elasticity of demand for terrorism insurance (percent change in the amount of terrorism coverage purchased divided by the percent change in price) is –0.65 (Kunreuther, 2005, p. 178). While Wharton reports that its estimate is statistically different from zero, inadequate information is given to calculate a statistical confidence interval. Absent better information, we allow the demand elasticity to rise and fall 50 percent from this point estimate. The resulting range for this parameter is –0.975 to –0.325.

Expense Factor. The expense loading for underwriting and loss-adjustment expenses ranges from about 30 percent to 50 percent of premiums across a wide range

---

[8] A *captive insurer* is an entity formed primarily to insure or reinsure the risk of one or a small number of policyholders.

of insurance lines (Harrington and Niehaus, 2004, p. 146). Because of the long tail of the exceedance probability curve for terrorism risk, we apply this factor to the expected loss as opposed to the sum of expected loss and the cost of the capital needed to protect against the possibility that losses exceed expected losses. Converting from a proportion of premium (which includes expense) to a multiple of the expected loss results in an expense factor that ranges from 1.43 to 2.0.[9]

Cost of Capital. Using a variety of methods, Cummins and Phillips (2005) concluded that the cost of equity capital for property and liability insurers ranges from 12 to 21 percent. After netting out the risk-free return on capital (which insurers will be able to earn while they hold the capital), the net cost of capital to insurers ranges from about 7 to 16 percent. Ongoing work by Michel-Kerjan (2007) suggests that insurers' cost of capital could be in the 5 to 10 percent range, and Lakdawalla and Zanjani (unpublished, pp. 63–64) find net costs for catastrophe bonds in the 5 to 6 percent range. We thus chose 5 percent as the lower bound for the cost of capital that insurers secure to protect themselves against the possibility that actual losses exceed insured losses. Hedge funds often aim for returns of 20 to 30 percent (which drops to roughly 15 to 25 percent after the risk-free rate of return is deducted) and are already active in providing capital for terrorism-insurance markets. While 25 percent seems high relative to results of previous studies, we set the upper bound on the cost of capital at 25 percent to capture a wide range of possible values.

Insurer Copayment Over TRIA Program Cap. There was considerable disagreement among the insurers contacted during this study about whether insurers face any liability for insured losses that exceed the $100 billion TRIA cap. Some thought that the statute establishing the program very clearly stated that insurers would have no liability. Others believed that, despite the cap, insurers would be required to pay some fraction of insured losses that exceed $100 billion. They pointed out that the statute states that "Congress shall determine the procedures for and the source of any payments for such excess insured losses" (P.L. 107-297, §103[e][3]) and that Congress could conceivably impose taxes on insurers to cover the loss. They also questioned whether the statute fully preempts claims that might be made in state court against insurers that do not pay all insured losses from an attack. In particular, they pointed to the U.S. Department of the Treasury's comment (the treasury administers the TRIA program) during the rulemaking process that the preemption language in the statute does not apply to WC claims (U.S. Department of the Treasury, 2004).

In our analysis, we vary the proportion of insured losses insurers will pay over the cap from 0 to 75 percent. The lower end of the range reflects views that the cap is truly hard and that insurers have no responsibility for losses that exceed the cap. It seems unlikely that insurers would end up paying all insured losses even beyond the cap

---

[9] If the expense loading, as a fraction of combined expenses and expected loss is $x (0 < x < 1)$, then expenses are $\left(1/(1-x)\right)$ of the expected loss.

amount, but possible legal responsibility for all WC claims and the imposition taxes or court ruling in favor of policyholders for property claims means that the percentage could be substantial. If insured losses were divided equally between WC and property claims (which is the case for some of the attacks examined in this study) and insurers end up paying all WC claims and half the property claims over the cap, the insurance industry could end up paying our assumed maximum 75 percent of insured losses that go beyond the cap.

Parameterization of Different Government Interventions

Parameters that characterize government intervention in the terrorism insurance market are varied to project the change in take-up rate in each plausible future. Table B.2 shows the parameters that characterize TRIA and the expiration of TRIA (no government program). The parameters for TRIA reflect the 2007 version of TRIA, as described previously, with a cap that varies from hard to quite soft (insurers pay 75 percent of insured losses that exceed the cap). Expiration of TRIA eliminates the deductible, the government copayment for losses above the deductible, and the program cap. The take-up rate on property policies for NBCR attacks is not tied to the take-up rate for conventional attacks.

The four versions of enhancing TRIA change two parameters in the take-up rate model: the hardness of the cap and the deductible. Table B.3 summarizes the parameter changes for the four runs. Note that, for these versions of the take-up rate model, the conventional and NBCR take-up rates are constrained to be the same.

The parameterizations so far have all assumed that the existing TRIA cap varies from hard to very soft (from insurers paying no insured losses over $100 billion to insurers paying 75 percent of insured losses over $100 billion). We also project the

Table B.2
Parameters That TRIA Expiration Affects

Parameter	Symbol	2007 TRIA	No Government Program
Insurer deductible	D^i	20% of 2006 DEP in TRIA lines	0
Insurer copayment for losses above the insurer deductible and below the program cap (%)	c_1	15	100
Program cap	K	$100 billion	No cap
Insurer payments of insured losses over program cap	c_2^e	0 to 75%[a]	Not applicable[a]
NBCR take-up rate	γ_{NBCR}	Not tied to conventional take-up rate	Not tied to conventional take-up rate

[a] Insurers make required payments until available surplus exhausted.

Table B.3
Parameters That Are Affected by Expanding TRIA to Require Offering Conventional and NBCR-Attack Coverage

Parameter	Symbol	Existing Cap, 20% Deductible	Existing Cap, 7.5% Deductible	Hard Cap, 20% Deductible	Hard Cap, 7.5% Deductible
Insurer deductible	D^i	20% of 2006 DEP in TRIA lines	7.5% of 2006 DEP in TRIA lines 0	20% of 2006 DEP in TRIA lines	7.5% of 2006 DEP in TRIA lines
Insurer copayment below program cap (%)	c_1	15	15	15	15
Program cap	K	$100 billion	$100 billion	$100 billion	$100 billion
Insurer payments of insured losses over program cap	c_2^e	0 to 75%	0 to 75%	0	0
NBCR take-up rate	γ_{NBCR}	Not tied to conventional take-up rate	Not tied to conventional take-up rate	Set equal to conventional take-up rate	Set equal to conventional take-up rate

effect of expanding TRIA to require offering NBCR coverage if the existing cap is hard. Such projections are made by setting the insurer payments for insured losses over the program cap under TRIA to zero and simulating the effects of only the two hard-cap expansions of TRIA. (It does not seem plausible that the existing cap could be hard and then the cap under the expanded program would become soft.)

Calibration of the Model

We use the model to project the take-up rate on property policies for conventional attacks if TRIA were allowed to expire under the full experimental design of plausible futures described in Chapter Two. For some futures, the take-up rate does not decline at all from its level with TRIA, and, in others, it declines to very low levels. To analyze the effect of TRIA expiring and project the take-up rate for TRIA expanded to better address NBCR, we restrict our attention to those scenarios in which the change in take-up with and without TRIA is consistent with previous empirical studies on the effect of TRIA expiring. Doing so is a way of calibrating the model. It restricts attention to those sets of parameter values (e.g., the exceedance probability insurers use to determine the amount of capital they must hold to cover losses) that produce the change in take-up rate estimated by empirical studies. Our approach of the impact of TRIA expiring is thus based on existing empirical studies, not the predictions of the take-up model. We do use the model to predict the changes in take-up rate with the expanded version of TRIA. The model predicts the take-up rate for property policies for TRIA with NBCR coverage if one believes that allowing TRIA to expire would cause the take-up rate for property policies for conventional attacks to fall by amounts consistent with existing studies.

Four studies provide some empirical evidence on how take-up rates might fall if TRIA were to expire. We summarize the results of each study and then draw on the results to set the range used in our analysis for the decline in take-up rate should TRIA expire.

In 2005, Aon analyzed insurance policies with terms that extended beyond TRIA's original December 31, 2005, expiration date. Aon (2005, pp. 21–22) found that absolute or substantial coverage sublimits for terrorism coverage were being imposed on 70 percent or more of available capacity for insurance with terms beyond December 31 (or that 30 percent of capacity would include terrorism coverage even if TRIA expired). As discussed previously, Aon estimated that take-up rates under TRIA were roughly 60 percent, so a decline to 30 percent represents a 50 percent decline in take-up rate.

In the last wave of its survey of insurers, the U.S. Department of the Treasury (2005, p. 76) asked whether policies written or renewed in the first two months of 2005 incorporated coverage for international terrorism in 2006 that was roughly similar to the coverage that those policies provided in 2005. (Insurers were requested to restrict their attention to policies that did have terrorism coverage in 2005.) Given that policies typically run for one year, the responses provide insight into the willingness of insurers to write terrorism coverage in the absence of TRIA. The survey results suggest that take-up rate for terrorism coverage would fall from between 38 percent and 50 percent.

David Cummins, a leading academic on insurance issues, inferred how the expiration of TRIA would affect take-up rates by comparing take-up rates for foreign acts of terrorism (covered by TRIA) with take-up rate for domestic, or so-called noncertified, acts of terrorism (not covered by TRIA). He noted that approximately 90 percent of insurers wrote coverage for foreign acts of terrorism in 2002 through 2004, while 40 percent wrote coverage for domestic attacks. Given that domestic terrorism events are generally viewed as less risky than foreign attacks, he concluded that, if TRIA expired, no more than 40 percent of insurers would offer coverage for foreign attacks (Cummins, 2006, p. 370). If this decline applies evenly across insurers of all sizes, take-up rates would be expected to fall by 56 percent.

A survey of commercial and multifamily mortgage lenders by the Mortgage Bankers Association provides a final piece of evidence about how take-up rates would decline if TRIA were to expire. Based on survey responses, the Mortgage Bankers Association (2004) projected that the proportion of loan balances covered by terrorism coverage would fall by 76 percent if TRIA were to expire.

Estimates of the percentage decline in take-up rate that would result if TRIA expired range from 38 to 76 percent in the four studies reviewed.[10] The spread in the estimates reflects both the different methods used and the many uncertainties underlying the issue. Substantial uncertainty remains over how the expiration of TRIA would affect take-up; to balance out the one estimate (from the Mortgage Bankers Association) that is substantially higher than the other three, we set the range over which the take-up rates falls from 25 to 75 percent.

Effect of Requiring Insurers to Offer Coverage for Conventional and NBCR Attacks on Demand for Coverage

The change in take-up rate if TRIA is modified to require NBCR offer is calculated by comparing the cost of terrorism coverage on property policies at the observed take-up rates for conventional and NBCR coverage (55 to 65 percent for conventional and 3 to 10 percent for NBCR), with the cost if the NBCR take-up rate rises to the same level as the take-up rate for conventional coverage under TRIA (55 to 65 percent). As illustrated in Figure B.1, this approach for calculating the change in take-up rate ignores the possibility of a downward-sloping supply curve. Consequently, our estimates will tend to understate the decline in take-up rate. Our assumption that the demand curve does not shift when NBCR coverage is added, however, will tend to cause our estimates to overstate the decline in take-up rate. The demand curve in Figure B.1 depicts demand for coverage for conventional terrorist attacks. It is reasonable to expect that policyholders would be willing to pay more for an insurance policy that covers both conventional and NBCR attacks and, thus, for the addition of NBCR coverage to shift the demand curve upward. While such a shift would mute the effect of increased prices on take-up, the question is by how much.

We think it plausible that the upward shift in the demand curve will not be large relative to the change in cost and, thus, that ignoring it will not have a substantial impact on our findings. First, previous work suggests that the demand for NBCR coverage is not great. A 2004 survey of policyholders by the U.S. Department of the Treasury found that only 3 percent of policyholders purchased coverage for NBCR attacks and that the primary reason policyholders gave for not purchasing NBCR coverage was that they believed that they were not at risk.[11] The widespread perception among policyholders that they are not at risk suggests that the inclusion of NBCR coverage would not lead to a substantial upward shift in the demand curve.

---

10 Studies on the impact of TRIA's expiration on take-up are all based on the expiration of the 2005 version of TRIA. However, as discussed previously in this appendix, we expect take-up rates under the 2005 and 2007 versions of TRIA to be comparable. Thus, the findings of these studies should apply to the expiration of the 2007 version of TRIA.

11 For those policyholders responding to the question, 62 percent reported that a major reason for not purchasing terrorism coverage was that they were not at risk (U.S. Department of the Treasury, 2005, pp. 106–107).

Second, the low take-up rate for NBCR coverage suggests that, for a large number of policyholders, the price of NBCR coverage is high relative to the perceived value of the coverage. Thus, it seems reasonable to expect the increase in price caused by adding NBCR coverage to conventional coverage would be large relative to the upward shift in demand caused by augmenting the coverage. Again, the implication is that ignoring the shift in demand is not likely to have a major impact on our findings.

APPENDIX C

Loss-Distribution Model

Insured Losses and Paying Parties

The loss-distribution model uses the total loss for different terrorist attacks from the RMS model, the government intervention structure, and the policyholder take-up from the take-up model to distribute losses among the insurance industry, uninsured loss, future policyholders, unpaid insured loss, and taxpayers.

The insured loss for any attack is estimated from the total loss, the property coverage take-up, and the fraction of property loss resulting from fire.[1] Following the discussion in Appendix B, the total insured loss is

$$I = L_{WC} + \left(\gamma + f\left(1-\gamma\right)\right)L_P.$$

The RMS model provides estimates of the fire losses for each type of attack. Table C.1 lists the fraction of losses resulting from fire for the six attack types considered in this study.

Table C.1
Fraction of Loss Resulting from Fire

Attack Type	Fraction of Loss Resulting from Fire, *f*
Nuclear bomb	0.18
Outdoor anthrax	0
Indoor chemical	0
Radiological bomb	0
10-ton car bomb	0.1
1-ton car bomb	0.1

[1] The take-up model uses estimates of the fraction of property loss resulting from fire to estimate insurer costs. The fire fraction must be applied in the loss-distribution step to actually allocate the loss among insured and uninsured parties.

In the remainder of this appendix, we describe how losses are distributed among the stakeholders under the different government interventions considered in this study. Note that our model includes the possibility for postevent government compensation of both uninsured and unpaid insured loss under all government interventions considered.

No Government Program

Insurance Industry Loss

The insurance industry loss is the insured loss less any unpaid insured loss. In a large disaster in which the insured loss exceeded the insurance industry surplus, some fraction of insurance claims would not be paid to insureds because the amount of available surplus, Q, limits the maximum amount the insurance industry can pay. In fact, many insurers would go bankrupt and be unable to pay claims long before the insured loss reached the industry surplus (e.g., Cummins et al., 2002), but we assume that the insurance industry would nonetheless pay these losses through guarantee funds.[2] The loss paid by the insurance industry, B, is then given by

$$B = \begin{cases} I & \text{for } I < Q \\ Q & \text{for } I \geq Q \end{cases}.$$

The surplus Q is one of the parameters whose value is varied to create the ensemble of plausible futures used in the RDM analysis. As shown in Table 2.3 in Chapter Two, our analysis considers a range of values for the available surplus, \$120 billion to \$240 billion, which is 33 percent above and below the current estimate of \$180 billion surplus for TRIA-relevant lines.[3]

---

[2] In developing the take-up rate model, we used data from ISO to determine the percentage of insurers that an attack would affect. This proportion is not relevant here, because guarantee funds can levy assessments on all insurers, not just those affected by an attack. Thus, in our analysis, neither the number of insurers bankrupted nor the percentage of insurers affected by an attack influences the insured loss paid by the insurance industry. Tapping guarantee funds could, in principle, allow the insurance industry to pay more than the available industry surplus. However, it seems likely that guarantee funds would be so heavily burdened by the time industry surplus is exhausted that relying on them for additional funds would be unrealistic. Finally, our model assumes that levies by guarantee funds would be paid out of insurer surplus. It is possible, however, that insurers may be allowed by state regulators to pass levies on to their policyholders. If such were the case, future policyholders may pay some of the losses we assign to insurers. We do not run scenarios that include such a possibility.

[3] In 2006, the President's Working Group on Financial Markets estimated that the DEP in TRIA lines for 2006 would be \$182 billion (PWG, 2006, p. 35). Net earned premium (which nets out insurer purchases of reinsurance) was \$436 billion in 2006, suggesting that the TRIA lines accounted for approximately 42 percent of premiums in the property-casualty industry. The property-casualty industry finished 2006 with \$486 billion in

Uninsured Loss

Uninsured loss is all losses not insured by commercial insurers less any ex post government compensation. Uninsured loss, U, is given by $U = (L - I)(1 - g_u)$, where g_u is the fraction of the original uninsured loss compensated by the government.

Unpaid Insured Loss

Insured victims are responsible for all insured loss not paid by commercial insurers less any ex post government compensation. Unpaid insured loss, V, is given by

$$V = \begin{cases} 0 & I < Q \\ (I - Q)(1 - g_V) & I \geq Q \end{cases},$$

where g_V is the fraction of the unpaid insured loss compensated by the government.

Taxpayer Cost

With no government intervention, taxpayers are responsible only for any ex post government compensation for the uninsured and unpaid insured loss. Taxpayer cost, T, is given by

$$T = \begin{cases} g_U(L - I) & I < Q \\ g_U(L - I) + g_V(I - Q) & I \geq Q \end{cases}.$$

TRIA

Insurance Industry Loss

Under TRIA, the insurance industry pays a portion of the insured loss determined by an insurer deductible and a copayment. Each insurer has a fixed deductible (a fraction of its annual DEP for the previous year in specified commercial property-casualty insurance lines). The aggregate industry payment under TRIA therefore depends on the number and sizes of insurers suffering losses and on the distribution of losses among those insurers. In general, the more widely spread the losses are among insurers, the greater the industry share.

---

surplus (Hartwig, 2007). The surplus attributable to the TRIA lines thus would be on the order of \$180 billion (0.42 × \$486 billion).

Modeling the insurance industry share is complicated by the fact that there is no obvious way to predict the distribution of annual terrorism losses among insurers.[4] The relationship developed in Appendix B is used to project the percentage of the premium base (β^1) affected by the attack. The applicable annual industrywide TRIA deductible, D, for any size insured loss can then be computed by multiplying the total aggregate industry deductible, *DA* (20 percent of DEP from the prior year in specified TRIA lines, which gives a 2007 aggregate deductible of $36 billion), by the market share, β^1: $D = D_A\beta^1$.

Before using this deductible to calculate insurer payments following a terrorist attack, it is useful to compare our method for approximating the insurer deductible with the approach used in a recent study by Kunreuther and Michel-Kerjan (2006).

Kunreuther and Michel-Kerjan (2006) used a micro approach for WC coverage that is based on the characteristics of the WC market in the state in which the attack occurs. They distributed property losses across insurers according to the nationwide market share. Our approach is equivalent to distributing losses across insurers in proportion to combined WC and property premiums for policies nationwide.

The two approaches produce somewhat different estimates of the insurer payments following a terrorist attack. For example, Kunreuther and Michel-Kerjan (2006) used WC market share data by state for the nation's 451 largest insurers to compare insurance industry losses for a $25 billion bomb attack that occurred in New York City, Los Angeles, or Houston. They found that the industry payments under TRIA following the same attack in each of the three cities varied by about 10 percent, from $13.1 billion to $14.5 billion. Our estimate of the industry payments using a single, national market-share model (for WC and property lines subject to TRIA combined) for the equivalent attack and same TRIA parameters and take-up is $17.5 billion (or about 25 percent larger than the midpoint of the Kunreuther–Michel-Kerjan estimate). The Kunreuther–Michel-Kerjan estimate may be lower than ours because companies that write exclusively WC coverage dominate the WC markets in the states they examined. Because the TRIA deductible is based on premiums in WC and commercial property lines, companies writing only WC coverage will have lower TRIA deductibles for a given WC loss than will companies writing a mix of WC and commercial property coverage.

The differences between the two estimates for this attack will likely decline as the Kunreuther–Michel-Kerjan (2006) approach is extended to more cities. Also, the difference between the two estimates will decline and ultimately reverse as the size

---

[4] Two end members bracket the industry share. If a single small insurer incurred all the insured loss from an attack, the deductible would be very small, and the industry share would essentially consist of the copayment. If the insured loss were distributed among many large insurers, each insurer's loss could be less than its deductible, in which case the industry would pay the entire insured loss until the insured loss exceeded the aggregate industry deductible (the sum of individual deductibles for all terrorism insurers in the nation). The industry share will fall somewhere between these two extremes.

of the attack grows because of the relationship we assume between market share of insurers incurring losses and loss size. We use the ISO model, in which market share initially increases rapidly with loss size and then tapers off and is capped at 70 percent. Kunreuther and Michel-Kerjan assume that all loss is shared among all insurers in the market in proportion to their TRIA DEP (the relevant market is the state for WC losses and the nation for property losses). For a $40 billion loss, our estimate of insurer losses is between Kunreuther and Michel-Kerjan's estimates for the three cities, and, for a $100 billion loss, our estimate is 20 percent to 35 percent lower.

There is no way to predict exactly how losses will be distributed among insurers and, therefore, no way to know which relationship (the ISO model or a proportionate distribution of losses across insurers based on the proportion of premiums written in the relevant market) would better model reality. However, we note that, if the market includes some insurers whose coverage is concentrated in low-risk properties, then the market share affected by an attack may never reach 100 percent, and the ISO relationship may be more plausible than a proportionate allocation.

All things considered, it is not obvious which of the two approaches yields better estimates for expected insurer payments across all high-risk cities in the United States. Comparison of the two approaches warrants further examination.

Returning to our model of insurer payments, the amount the insurance industry pays toward the deductible, B_D, is then the lesser of the insured loss or D:

$$B_D = \begin{cases} I & \text{for } I < D \\ D & \text{for } I \geq D \end{cases}.$$

The industry copayment, I_C, is a fixed fraction of the insured loss above the deductible, with the maximum amount limited by the TRIA cap, K ($100 billion). B_C is thus expressed as

$$B_C = \begin{cases} c_1\left(I - B_D\right) & \text{for } I < K \\ c_1\left(K - B_D\right) & \text{for } I \geq K \end{cases},$$

where c_1 is the industry portion of the shared compensation feature of TRIA (15 percent in 2007). The nominal total industry share under TRIA, B_T, is the sum of the deductible and copayment: $B_T = B_D + B_C$.

For a total aggregate industry deductible of $36 billion, B_T is equal to the insured loss for insured losses up to about $22 billion. Therefore, under TRIA, the insurance industry is responsible for all losses up to $22 billion.

In our analysis, we allow for the possibility that the insurance industry will be responsible for some fraction of the losses above the program cap. This amount, B_K, is given by

$$B_K = \begin{cases} 0 & \text{for } I \leq K \\ c_2(I-K) & \text{for } I > K \end{cases},$$

where c_2 is the fraction of insured losses above the program cap that the industry is called on to pay. The industry copayment for losses above the cap is one of the parameters that we vary to create the ensemble of plausible futures used in the RDM analysis. TRIA has never been activated, so there is no precedent for what action, if any, would be taken to settle claims for insured losses above the cap. In our analysis, we allow for both government and industry to compensate these losses. The government contribution is included in g_V (see below). As shown in Table 2.3 in Chapter Two, we allow the industry payment for losses above the cap to vary from 0 percent to 75 percent.

The nominal insurance industry loss, B_N, is thus given as $B_N = B_D + B_C + B_K$.

Finally, as in the no-government-program case, some fraction of the insurance industry loss may not be paid, because the industry surplus will be exhausted. The total industry loss is therefore given by

$$B = \begin{cases} B_N & \text{for } B_N < Q \\ Q & \text{for } B_N \geq Q \end{cases}.$$

Uninsured Losses

The uninsured share is all loss not insured by commercial insurers less any ex post government compensation. Uninsured losses are given by $U = (L - I)(1 - g_U)$.

Future Policyholder Losses

Under TRIA, the government reimburses each insurer for its loss above its deductible and copayment; then, it recoups all or part of that reimbursement through a surcharge levied on all commercial property-casualty insurers in the country, whether or not they had purchased terrorism insurance. The amount recouped is governed by the "insurance marketplace aggregate retention amount," R ($27.5 billion in 2007). The future policyholder surcharge amount, S, is given by

$$S = \begin{cases} I - B_T & \text{for } I \leq R \\ R - B_T & \text{for } I > R \text{ and } B_T \leq R. \\ 0 & \text{for } B_T > R \end{cases}$$

For $I \le R$, S equals the federal reimbursement and, therefore, the entire federal payout is recouped. When $I > R$ and $B_T \le R$, the federal payout is only partially recouped, and, if $B_T > R$, none of the federal payout is recouped.

The government recoupment is one of the parameters whose value we vary for our RDM analysis. The government has never had the occasion to recoup payments under TRIA, so there is no precedent to guide our choice for this parameter. As shown in Table 2.3 in Chapter Two, we thus assume that it can range from \$27.5 billion to \$100 billion, its minimum to maximum values under the law.

Unpaid Insured Loss

Under TRIA, unpaid insured loss could arise from (1) insurance industry responsibilities not paid because of surplus limitations or (2) insured loss above the program cap that is not compensated by ex post industry or government compensation. Any industry payments for losses above the cap are included in B. Therefore, the unpaid insured loss is given by

$$V = \begin{cases} (B_T - B)(1 - g_V) & \text{for } I \le K \\ (B_T - B + I - K)(1 - g_V) & \text{for } I > K \end{cases}.$$

Taxpayer Cost

Under TRIA, taxpayers are responsible for all federal outlays not recouped through the commercial policyholder surcharge plus any ex post compensation for uninsured and unpaid insured loss. Taxpayer cost is given by

$$T = \begin{cases} I - B_T - S + g_U(L - I) + g_V(B_T - B) & \text{for } I \le K \\ K - B_T - S + g_U(L - I) + g_V(B_T - B + I - K) & \text{for } I > K \end{cases}.$$

TRIA with NBCR-Attack Coverage

Losses are distributed across the different stakeholders using the same methods as described for TRIA. The only difference is the value of the take-up rate.

APPENDIX D

Identifying Key Factors Driving Trade-Offs Between Interventions

Following Lempert, Groves, et al. (2006) and Groves and Lempert (2007), we identify the key uncertain input factors driving the trade-offs among the government interventions by applying Friedman and Fisher's (1999) "patient" rule induction method (PRIM) to the model-generated database representing the ensemble of plausible futures. PRIM is a data-mining algorithm designed to generate a set of low-dimensional "boxes" in high-dimensional data containing regions in which the value of a particular function is large (or small) compared to its value outside these boxes. PRIM is particularly useful for identifying the key factors driving the trade-offs among the alternative interventions, because it aims to optimize both the classification accuracy of the boxes (the percentage of large or small function values they contain) and the interpretability of the boxes (the simplicity of the rules needed to define them).

We implement PRIM using publicly available software[1] that inputs a data set (which can be the output of a model run over many combinations of input values) and a criterion for interesting cases. The algorithm outputs descriptions of several alternative low-dimensional regions, or boxes, that contain a high density and span a high proportion of the interesting cases. We define a record of interest in the ensemble of plausible futures as one in which one of the measures of merit for a government intervention exceeds some threshold value. We also consider cases of interest in which the measure of merit for one intervention, e.g., cost to taxpayers under TRIA, exceeds a measure for another intervention, e.g., cost to taxpayers if TRIA expires. With such definitions, PRIM naturally generates two useful quantitative measures of merit for the resulting clusters—the coverage and density of total states of interest captured.

PRIM is inherently interactive, in the sense that each run suggests several alternative clusters among which the user is asked to choose. In particular, the coverage and density measures are generally inversely correlated, since a larger cluster may include a lower density of high-value data. The PRIM software thus presents the user with trade-off curves that display clusters with different combinations of coverage and density. These clusters often differ in the number and identity of the driving forces that define

---

1 SuperGEM™ is an implementation of PRIM.

them. Users then choose the cluster with the desired density/coverage trade-off and interpretability—that is, the one whose particular set of defining driving forces makes it meaningful to the user. After choosing a cluster, the records within it are removed from the database and PRIM can be run again on the remaining records to produce additional clusters.

Figure D.1 suggests the trade-offs typically offered by the PRIM algorithm. Imagine a data set with six records of interest (shown by solid dots) and a total of 25 records. The smaller box contains only points of interest (100 percent density) but captures only four of the six such points (67 percent coverage). The larger box captures all six points of interest (100 percent coverage) but contains three additional, unwanted points (67 percent density). In this example, the same two factors, represented by the axes of Figure D.1, describe both boxes. In general, this need not be the case for the clusters suggested by PRIM.

Table D.1 shows two clusters that explain the results shown in Chapter Three and are generated by applying PRIM to the 4,568-case data set representing the performance of TRIA in conventional terrorist attacks. Both these clusters aim to identify the key factors that cause costs to taxpayers under TRIA to exceed $10 billion. The cluster in the center column suggests that attacks with total losses larger than $38

Figure D.1
Coverage and Density Trade-Offs Offered by PRIM

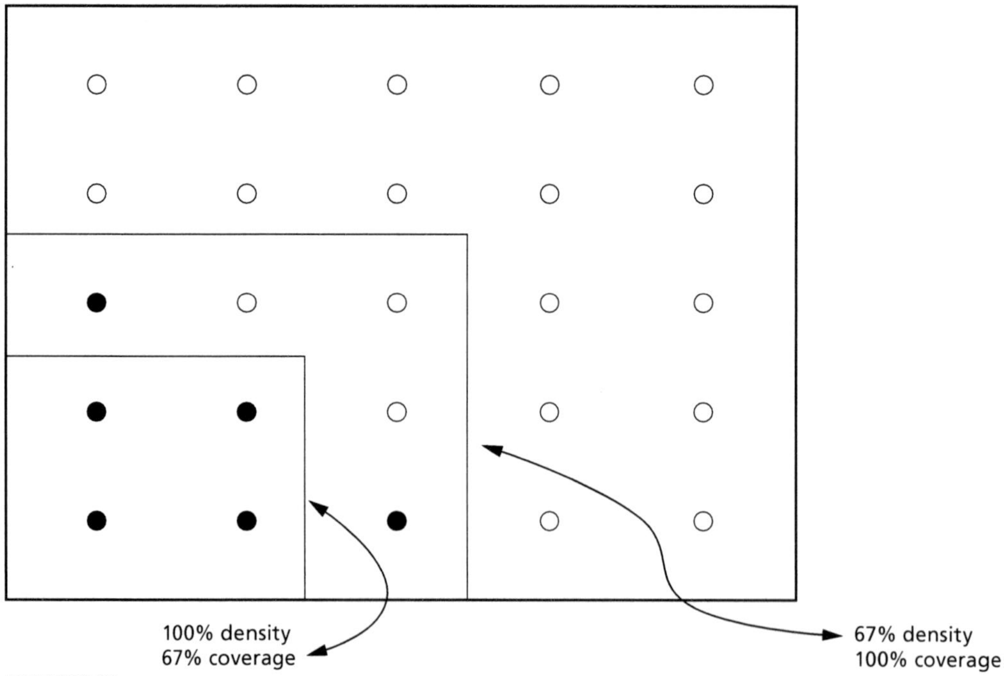

RAND *MG679-D.1*

Table D.1
PRIM-Generated Clusters Explaining the 964/4,568 = 21 Percent of Cases in Which TRIA Imposes High Costs (More Than $10 Billion) on Taxpayers

Ranges	Attack Causes >$38 Billion in Damage (%)	Attack Causes >$45 Billion in Damage (%)
Coverage	100	87
Density	85	100

billion explain all the cases with high taxpayer cost under TRIA, and all but 15 percent of such cases occur when losses exceed $38 billion. The cluster in the right column suggests that high taxpayer cost under TRIA occurs in all the cases with attacks larger than $45 billion, but 13 percent of cases in which TRIA imposes high costs have smaller attacks. The relatively narrow range of attack sizes between these two clusters suggests that the attack size totally dominates the effect of all the other uncertain parameters in Table 2.3 in Chapter Two in explaining the causes of high taxpayer cost under TRIA.

Table D.2 shows two PRIM clusters that explain more results shown in Chapter Three. Both these clusters aim to identify the key factors that cause the fraction of unpaid insured loss to exceed 20 percent if TRIA expires. The cluster in the center column suggests that a postattack government decision to compensate less than 55 percent of uninsured losses explains all the cases with a high fraction of unpaid insured loss, though only 70 percent of such cases have such a high level of unpaid insured loss. The cluster in the right column suggests that a high fraction of unpaid insured loss occurs in all the cases with postattack government compensation less than 21 percent, but 55 percent of such cases have larger government compensation. The relatively wide range of government compensation between these two clusters suggests that many of the other uncertain parameters in Table 2.3 in Chapter Two can, in some cases, be important in causing a high fraction of unpaid insured loss if TRIA expires.

Table D.3 shows two PRIM clusters that explain the additional results shown in Chapter Three. Both these clusters aim to identify the key factors that cause costs to taxpayers under TRIA to exceed the costs to taxpayers if TRIA expires. The cluster in

Table D.2
PRIM-Generated Clusters Explaining the 2,348/4,568 = 51 Percent of Cases with a High Fraction (More Than 20 Percent) of Unpaid Insured Loss If TRIA Expires

Ranges	Government Compensation to Uninsured <55% (%)	Government Compensation to Uninsured <21% (%)
Coverage	100	45
Density	70	100

Table D.3
PRIM-Generated Clusters Explaining the 1,177/4,568 = 26 Percent of Cases with Taxpayer Cost Under TRIA Higher Than That If TRIA Expires

Ranges	Attack Causes >$33 Billion in Damage (%)	Attack Causes >$40 Billion in Damage (%)
Coverage	100	88
Density	91	100

the center column suggests that attacks with total losses larger than $33 billion explain all the cases with higher taxpayer cost under TRIA, and all but 9 percent of such cases occur when losses exceed $33 billion. The cluster on the right suggests that TRIA costs taxpayers more in all the cases with attacks larger than $40 billion, but 12 percent of cases in which TRIA imposes higher costs have smaller attacks. The relatively narrow range of attack sizes between these two clusters suggests that the attack size totally dominates the effect of all the other uncertain parameters in Table 2.3 in Chapter Two in explaining the cases in which TRIA costs the taxpayer more than letting TRIA expire.

APPENDIX E

Calculating Expected Losses with Multiple Probability Distributions

As the results in Chapters Three and Four make clear, the size of a terrorist attack, as measured in total property and WC losses, plays an important role in the distribution of those losses among taxpayers, property owners, and the insurance industry. Any information on the likelihood of different-sized attacks would thus prove very useful. While estimates of these probability distributions exist, they are unavoidably deeply uncertain because our knowledge of and experience with such large attacks remains extremely limited. Considering sets of plausible probability distributions provides one common means of characterizing this type of uncertainty (Smithson, 1989). This appendix thus describes how we calculate the expected value of various outcome measures from our simulation model, contingent on a range of plausible distributions describing the likelihood of large terrorist attacks.

As described in Appendix A, RMS provides estimates of the probability of terrorist attacks of various sizes. The firm creates these distributions from elicitations with terrorism experts gathered for this purpose at semiannual conferences. In this study, we used these RMS distributions as the base from which we construct a set of distributions.

The left and right panels in Figure E.1 show the 2006 RMS-estimated probability distribution for losses from conventional and NBCR attacks, respectively (solid bars). These distributions suggest, not surprisingly, that any NBCR attack is much less likely than a conventional attack, but that tails of the NBCR distribution include much larger attacks than the conventional distribution (nearly $2 trillion, as compared to $70 billion).

The analysis in Chapter Three (also see Appendix D) suggests that $40 billion represents a key threshold loss when considering the costs to taxpayers under alternative government interventions. For attacks with more than $40 billion in losses, TRIA almost always costs taxpayers more than having no government program in place, while the reverse is true for attacks with less than $40 billion in losses. We thus create a set of plausible probability distributions by reweighting the tails of the RMS distributions for losses greater than $40 billion. Figure E.1 shows the tails of the conventional and NBCR distributions with 10 times (gray bars) and with one-tenth (white bars) the weight of the original RMS distributions. In each case, we also reweight that portion

Figure E.1
Probability Distributions for Conventional and NBCR Attacks

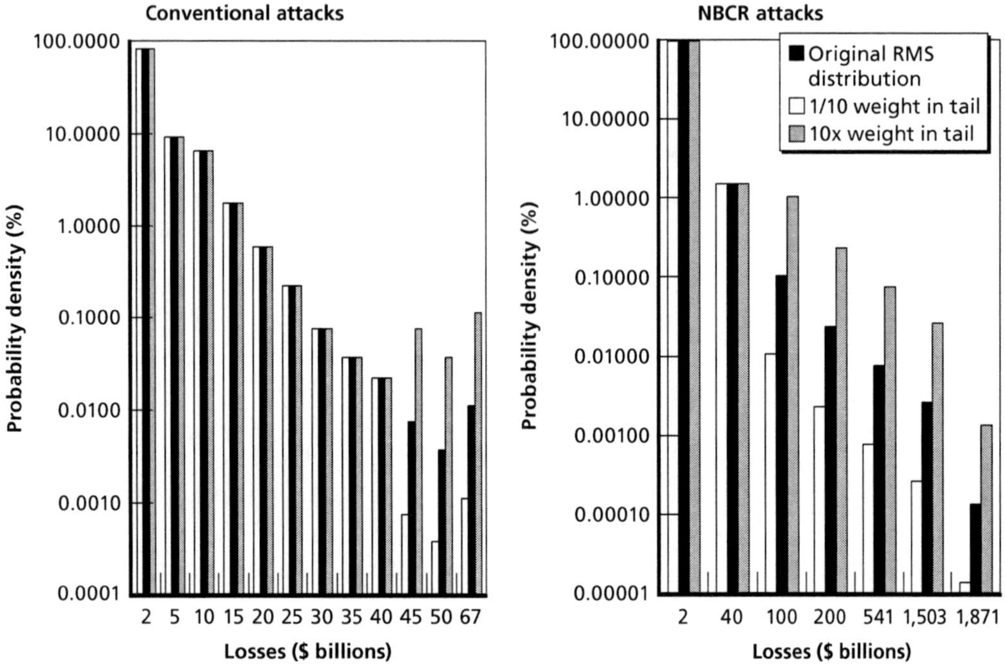

RAND *MG679-E.1*

of the distribution for losses less than $40 billion so that the total integrates to unity. However, for both the conventional and NBCR distributions, the probability of an attack with more than $40 billion in losses is sufficiently small so that reweighting below this value is difficult to detect in the figure.

We now calculate the expected value of various performance measures using the set of distributions such as those shown in Figure E.1. Table E.1 shows the expected value of the change in taxpayer cost if TRIA expires—that is, the expected taxpayer cost if TRIA expires less the expected taxpayer cost with TRIA, as a function of the fraction of uninsured and unpaid insured loss that the government compensates after an attack and of the odds of a large attack relative to those in the RMS model. These odds are measured by the ratio of the weight in the tail of the distribution used to calculate the expected value of taxpayer losses and the weight in the tail of the RMS distribution. An odds ratio of 1 indicates that we calculate the expected values using the RMS distribution, an odds ratio of 10 indicates that we calculate expected values using a distribution whose tails for losses greater than $40 billion have 10 times the weight of that in the RMS distribution (the gray bars in Figure E.1), and odds of 0.1 indicate that we use a distribution whose tails for losses greater than $40 billion have one-tenth the weight of that in the RMS distribution (the white bars in Figure E.1).

Table E.1
Change in Expected Cost to Taxpayers If TRIA Expires (Cost with TRIA Less Cost Without TRIA) ($ Billions)

Government Compensation for Uninsured Losses and Unpaid Insured Losses	Ratio of Odds of a Large Attack to Odds of a Large Attack in RMS Model				
	0.1	1	10	50	100
A. Conventional attacks					
0	0.001	0.004	0.033	0.161	0.318
25	−0.077	−0.074	−0.048	0.065	0.205
50	−0.155	−0.153	−0.130	−0.031	0.091
75	−0.233	−0.231	−0.212	−0.126	−0.022
B. NBCR attacks					
0	0.001	0.012	0.117	0.555	1.041
25	0.001	0.012	0.118	0.559	1.050
50	0.001	0.012	0.119	0.563	1.058
75	0.001	0.012	0.120	0.568	1.066
C. All attacks					
0	0.002	0.016	0.150	0.716	1.359
25	−0.076	−0.062	0.070	0.624	1.254
50	−0.154	−0.141	−0.011	0.533	1.149
75	−0.232	−0.219	−0.092	0.441	1.044

The results in Table E.1 include both conventional and NBCR attacks. We calculated the expected losses separately for conventional and NBCR attacks, using the distributions in the left and right panels of Figure E.1, and then added the results.[1] Note that expected taxpayer cost is higher with TRIA than without TRIA for NBCR attacks for all values of the parameters examined. The results in Table E.1 are used in Figure 3.8 in Chapter Three.

Table E.2 shows similar expected value calculations for the change in expected costs to taxpayers for TRIA modified to include NBCR with low deductible and a hard cap compared to the expected cost to taxpayers under the current TRIA. Now, expected taxpayer cost is lower for TRIA with NBCR coverage than for TRIA for NBCR attacks for some parameter values (the difference between expected taxpayer

[1] The distributions of NBCR attacks and conventional attacks are likely correlated, but the expected value of the sum of two random variables is the sum of the expected values of each random variable, regardless of the correlation between the two variables.

Table E.2
Change in Expected Cost to Taxpayers for TRIA with NBCR-Attack Coverage, Low Deductible, and Hard Cap Versus TRIA (Cost Under TRIA with NBCR-Attack Coverage Less Cost Under TRIA) ($ Billions)

Government Compensation for Uninsured Losses and Unpaid Insured Losses	Ratio of Odds of a Large Attack to Odds of a Large Attack in RMS Model				
	0.1	1	10	50	100
A. Conventional attacks					
0	0.001	0.001	0.004	0.020	0.039
25	−0.013	−0.013	−0.010	0.001	0.015
50	−0.027	−0.027	−0.025	−0.018	−0.009
75	−0.037	−0.037	−0.036	−0.032	−0.028
B. NBCR attacks					
0	0.007	0.065	0.637	3.016	5.661
25	−0.021	0.019	0.404	2.008	3.791
50	−0.058	−0.038	0.161	0.990	1.912
75	−0.081	−0.079	−0.067	−0.013	0.046
C. All attacks					
0	0.008	0.066	0.641	3.036	5.700
25	−0.034	0.005	0.394	2.009	3.806
50	−0.085	−0.065	0.136	0.973	1.903
75	−0.118	−0.116	−0.103	−0.046	0.018

cost for TRIA with NBCR coverage and TRIA is negative). Figure 4.9 in Chapter Four uses the results in Table E.2.

Table E.3 shows similar expected value calculations for the change in expected costs to taxpayers for TRIA modified to include NBCR with a low deductible and a hard cap compared to the expected cost to taxpayers if TRIA expires.

Table E.3
Change in Expected Cost to Taxpayers for Low Deductible, Hard Cap NBCR-Attack Offer Versus No Government Program (Cost Under TRIA with NBCR-Attack Coverage Less Cost Without TRIA) ($ Billions)

Government Compensation for Uninsured Losses and Unpaid Insured Losses	Ratio of Odds of a Large Attack to Odds of a Large Attack in RMS Model				
	0.1	1	10	50	100
A. Conventional attacks					
0	0.002	0.005	0.038	0.181	0.357
25	−0.098	−0.095	−0.067	0.058	0.212
50	−0.186	−0.184	−0.159	−0.052	0.079
75	−0.270	−0.268	−0.247	−0.159	−0.050
B. NBCR attacks					
0	0.008	0.077	0.754	3.571	6.703
25	−0.019	0.030	0.522	2.567	4.840
50	−0.057	−0.026	0.280	1.554	2.970
75	−0.079	−0.067	0.053	0.555	1.112
C. All attacks					
0	0.010	0.082	0.791	3.751	7.059
25	−0.118	−0.065	0.455	2.625	5.052
50	−0.243	−0.209	0.121	1.501	3.048
75	−0.349	−0.335	−0.194	0.396	1.062

References

Alesch, Daniel J., James N. Holly, Elliott Mittler, and Robert Nagy, *Organizations at Risk: What Happens When Small Businesses and Not-for-Profits Encounter Natural Disasters*, Green Bay, Wisc.: Public Entity Risk Institute, 2001.

Aon Corporation, "Property Terrorism Update: TRIA in the Balance," October 2005. As of May 9, 2007: http://www.aon.com/about/publications/issues/property_terrorism_update-oct2005.jsp

———, *Response to U.S. Treasury and President's Working Group: Terrorism Insurance*, April 21, 2006.

Bankes, Steve, "Exploratory Modeling for Policy Analysis," *Operations Research*, Vol. 41, No. 3, May 1993, pp. 435–449.

Barker, David, "Terrorism Insurance Subsidies and Social Welfare," *Journal of Urban Economics*, Vol. 54, No. 2, September 2003, pp. 328–338.

Ben-Haim, Yakov, *Information-Gap Decision Theory: Decisions Under Severe Uncertainty*, San Diego, Calif.: Academic Press, 2001.

Brown, Jeffrey R., J. David Cummins, Christopher M. Lewis, and Ran Wei, "An Empirical Analysis of the Economic Impact of Federal Terrorism Reinsurance," *Journal of Monetary Economics*, Vol. 51, No. 5, July 2004, pp. 861–898.

Carroll, Stephen J., Tom LaTourrette, Brian G. Chow, Gregory S. Jones, and Craig Martin, *Distribution of Losses from Large Terrorist Attacks Under the Terrorism Risk Insurance Act*, Santa Monica, Calif.: RAND Corporation, MG-427-CTRMP, 2005. As of September 6, 2007: http://www.rand.org/pubs/monographs/MG427/

Chalk, Peter, Bruce Hoffman, Robert T. Reville, and Anna-Britt Kasupski, *Trends in Terrorism: Threats to the United States and the Future of the Terrorism Risk Insurance Act*, Santa Monica, Calif.: RAND Corporation, MG-393-CTRMP, 2005. As of September 6, 2007: http://www.rand.org/pubs/monographs/MG393/

Code of Federal Regulations, Title 31, Part 50.5, Paragraph (d)(1), State Licensed or Admitted Insurers, July 1, 2007. As of September 7, 2007: http://a257.g.akamaitech.net/7/257/2422/03jul20071500/edocket.access.gpo.gov/cfr_2007/julqtr/pdf/31cfr50.5.pdf

Cummins, J. David, "Statistical and Financial Models of Insurance Pricing and the Insurance Firm," *Journal of Risk and Insurance*, Vol. 58, No. 2, June 1991, pp. 261–302.

———, "Should the Government Provide Insurance for Catastrophes?" *Review* (Federal Reserve Bank of St. Louis), Vol. 88, No. 4, July–August 2006, pp. 337–385. As of September 6, 2007: http://research.stlouisfed.org/publications/review/06/07/Cummins.pdf

Cummins, J. David, Neil Doherty, and Anita Lo, "Can Insurers Pay for the 'Big One'? Measuring the Capacity of the Insurance Market to Respond to Catastrophic Losses," *Journal of Banking and Finance*, Vol. 26, Nos. 2–3, March 2002, pp. 557–583.

Cummins, J. David, and Richard D. Phillips, "Estimating the Cost of Equity Capital for Property-Liability Insurers," *Journal of Risk and Insurance*, Vol. 72, No. 3, September 2005, pp. 441–478.

Dewar, James A., *Assumption-Based Planning: A Tool for Reducing Avoidable Surprises*, Cambridge and New York: Cambridge University Press, 2002.

Dixon, Lloyd, John Arlington, Stephen J. Carroll, Darius N. Lakdawalla, Robert T. Reville, and David Adamson, *Issues and Options for Government Intervention in the Market for Terrorism Insurance*, Santa Monica, Calif.: RAND Corporation, OP-135-ICJ, 2004. As of September 6, 2007: http://www.rand.org/pubs/occasional_papers/OP135/

Dixon, Lloyd, Noreen Clancy, Seth A. Seabury, and Adrian Overton, *The National Flood Insurance Program's Market Penetration Rate: Estimates and Policy Implications*, Santa Monica, Calif.: RAND Corporation, TR-300-FEMA, 2006. As of September 6, 2007: http://www.rand.org/pubs/technical_reports/TR300/

Dixon, Lloyd, Robert Lempert, Tom LaTourrette, Robert T. Reville, and Paul Steinberg, *Trade-Offs Among Alternative Government Interventions in the Market for Terrorism Insurance: Interim Results*, Santa Monica, Calif.: RAND Corporation, DB-525-CTRMP, 2007. As of September 6, 2007: http://www.rand.org/pubs/documented_briefings/DB525/

Dixon, Lloyd, and Rachel Kaganoff Stern, *Compensation for Losses from the 9/11 Attacks*, Santa Monica, Calif.: RAND Corporation, MG-264-ICJ, 2004. As of September 6, 2007: http://www.rand.org/pubs/monographs/MG264/

Dong, Weimin, Haresh Shah, and Felix Wong, "A Rational Approach to Pricing of Catastrophe Insurance," *Journal of Risk and Uncertainty*, Vol. 12, Nos. 2–3, May 1996, pp. 201–218.

Friedman, Jerome H., and Nicholas I. Fisher, "Bump Hunting in High-Dimensional Data," *Statistics and Computing*, Vol. 9, No. 2, April 1999, pp. 123–143.

GAO—*see* U.S. General Accounting Office.

Groves, David G., and Robert J. Lempert, "A New Analytic Method for Finding Policy-Relevant Scenarios," *Global Environmental Change*, Vol. 17, No. 1, February 2007, pp. 73–85.

Harrington, Scott E., and Greg Niehaus, *Risk Management and Insurance*, 2nd ed., Boston, Mass.: McGraw-Hill, 2004.

Hartwig, Robert P., "2006: Year End Results," Insurance Information Institute, New York, April 18, 2007. As of May 10, 2007: http://www.iii.org/media/industry/financials/2006yearend/

Hubbard, R. Glenn, Bruce Deal, and Peter Hess, "The Economic Effects of Federal Participation in Terrorism Risk," *Risk Analysis and Insurance Review*, Vol. 8, No. 2, September 2005, pp. 177–209.

Insurance Services Office, "Terrorism Loss Costs and Complementary Rating Rules Being Filed for Commercial Property in Response to the Terrorism Risk Insurance Act of 2002; Rule Updated for Other Property Lines," Jersey City, N.J.: Insurance Services Office, ISO circular, December 23, 2002.

ISO—*see* Insurance Services Office.

Jaffee, Dwight M., and Thomas Russell, *Should Governments Support the Private Terrorism Insurance Market?* paper prepared for August 2005 World Risk and Insurance Economics Congress conference, Salt Lake City, Utah, June 28, 2005. As of September 6, 2007: http://faculty.haas.berkeley.edu/jaffee/Papers/DJTRSLCPaper.pdf

Kunreuther, Howard, *TRIA and Beyond: Terrorism Risk Financing in the U.S.*, Philadelphia, Pa.: Insurance and Risk Management Department, Wharton School, University of Pennsylvania, August 1, 2005. As of May 9, 2007: http://knowledge.wharton.upenn.edu/paper.cfm?paperid=1310

Kunreuther, Howard, and Erwann Michel-Kerjan, *Looking Beyond TRIA: A Clinical Examination of Potential Terrorism Loss Sharing*, Cambridge, Mass.: National Bureau of Economic Research, NBER working paper 12069, March 2006. As of May 9, 2007: http://www.nber.org/papers/W12069

LaTourrette, Tom, David R. Howell, David E. Mosher, and John MacDonald, *Reducing Terrorism Risk at Shopping Centers: An Analysis of Potential Security Options*, Santa Monica, Calif.: RAND Corporation, TR-401, 2006. As of May 9, 2007: http://www.rand.org/pubs/technical_reports/TR401/

Lakdawalla, Darius N., and George Zanjani, *Catastrophe Bonds, Reinsurance, and the Optimal Collateralization of Risk Transfer*, Santa Monica, Calif.: RAND Corporation, unpublished research.

Lempert, Robert J., and Myles Collins, "Managing the Risk of Uncertain Threshold Response: Comparison of Robust, Optimum, and Precautionary Approaches," *Risk Analysis*, forthcoming.

Lempert, Robert J., David G. Groves, Steven W. Popper, and Steve C. Bankes, "A General, Analytic Method for Generating Robust Strategies and Narrative Scenarios," *Management Science*, Vol. 52, No. 4, April 2006, pp. 514–528.

Lempert, Robert J., and Steven W. Popper, "High-Performance Government in an Uncertain World," in Robert Klitgaard and Paul C. Light, eds., *High-Performance Government: Structure, Leadership, Incentives*, Santa Monica, Calif.: RAND Corporation, MG-256-PRGS, 2005, pp. 113–138. As of May 8, 2007: http://www.rand.org/pubs/monographs/MG256/

Lempert, Robert J., Steven W. Popper, and Steven C. Bankes, *Shaping the Next One Hundred Years: New Methods for Quantitative, Long-Term Policy Analysis*, Santa Monica, Calif.: RAND Corporation, MR-1626-RPC, 2003. As of September 6, 2007: http://www.rand.org/pubs/monograph_reports/MR1626/

Levin, Andrew T., and John C. Williams, "Robust Monetary Policy with Competing Reference Models," *Journal of Monetary Economics*, Vol. 50, No. 5, July 2003, pp. 945–975.

Marsh, *Marketwatch: Terrorism Insurance 2006*, New York: Marsh, 2006. As of September 10, 2007: http://global.marsh.com/news/articles/terrorism/documents/MarketwatchTerrorism2006.pdf

Michel-Kerjan, Erwann, managing director, Center for Risk Management and Decision Processes, Wharton School, University of Pennsylvania, comment on *Trade-Offs Among Alternative Government Interventions in the Market for Terrorism Insurance*, emailed to Lloyd Dixon, Robert Reville, and Tom LaTourrette, April 7, 2007.

Morgan, M. Granger, and Max Henrion, *Uncertainty: A Guide to Dealing with Uncertainty in Quantitative Risk and Policy Analysis*, Cambridge and New York: Cambridge University Press, 1990.

Mortgage Bankers Association, *Study of Terrorism Insurance, TRIA and the "Make-Available" Provision*, Washington D.C., 2004.

Myers, Stewart C., and James A. Read Jr., "Capital Allocation for Insurance Companies," *The Journal of Risk and Insurance,* Vol. 68, No. 4, December 2001, pp. 545–580.

Popper, Steven W., Robert J. Lempert, and Steven C. Bankes, "Shaping the Future," *Scientific American*, Vol. 292, No. 4, April 2005, pp. 66–71.

President's Working Group on Financial Markets, *Terrorism Risk Insurance: Report of the President's Working Group on Financial Markets*, September 2006. As of May 9, 2007:
http://www.treas.gov/offices/domestic-finance/financial-institution/terrorism-insurance/pdf/report.pdf

Public Law 107-297, Terrorism Risk Insurance Act of 2002, November 26, 2002. As of May 9, 2007:
http://frwebgate.access.gpo.gov/cgi-bin/getdoc.cgi?dbname=107_cong_public_laws&docid=f:publ297.107.pdf

Public Law 109-144, Terrorism Risk Insurance Extension Act of 2005, December 22, 2005. As of May 9, 2007:
http://www.treas.gov/offices/domestic-finance/financial-institution/terrorism-insurance/pdf/tria_pl-109-144.pdf

PWG—*see* President's Working Group on Financial Markets.

Risk Management Solutions, "RMS Home," undated homepage. As of May 8, 2007:
http://www.rms.com

———, *Managing Terrorism Risk*, Newark, Calif.: Risk Management Solutions, 2003. As of December 20, 2006:
http://www.rms.com/publications/terrorism_risk_modeling.pdf

RMS—*see* Risk Management Solutions.

Rosenhead, Jonathan, "Robustness Analysis: Keeping Your Options Open," in John Mingers and Jonathan Rosenhead, eds., *Rational Analysis for a Problematic World Revisited: Problem Structuring Methods for Complexity, Uncertainty and Conflict*, 2nd ed., Chichester and New York: Wiley, 2001, pp. 181–208.

Saltelli, A., K. Chan, and E. Marian Scott, *Sensitivity Analysis*, Chichester and New York: Wiley, 2000.

Saxton, Jim, *Economic Perspectives on Terrorism Insurance*, Washington, D.C.: U.S. Congress, Joint Economic Committee, May 2002. As of September 6, 2007:
http://www.house.gov/jec/terrorism/insur.pdf

SBA—*see* U.S. Small Business Administration.

Smithson, Michael, *Ignorance and Uncertainty: Emerging Paradigms*, New York: Springer-Verlag, 1989.

SuperGEM™ 1.0 with S/Splus, software. As of September 12, 2007:
http://stat.stanford.edu/~jhf/SuperGEM.html

Torregrosa, David, *Federal Terrorism Reinsurance: An Update*, Washington, D.C.: Congress of the United States, Congressional Budget Office, 2005. As of September 6, 2007:
http://purl.access.gpo.gov/GPO/LPS72661

———, *Federal Reinsurance for Terrorism Risks: Issues in Reauthorization*, Washington, D.C.: Congress of the United States, Congressional Budget Office, August 2007. As of September 6, 2007:
http://www.cbo.gov/ftpdocs/85xx/doc8520/08-02-TRIA.pdf

U.S. Department of the Treasury, "Terrorism Risk Insurance Program; Litigation Management," proposed rule, *Federal Register*, Vol. 69, No. 88, May 6, 2004.

———, Office of Economic Policy, *Assessment: The Terrorism Risk Insurance Act of 2002*, Washington, D.C.: U.S. Department of the Treasury, Office of Economic Policy, June 30, 2005. As of May 9, 2007:
http://www.treasury.gov/press/releases/reports/063005%20tria%20study.pdf

———, Office of Domestic Finance, "Terrorism Risk Insurance Program," Web page, last modified March 23, 2006. As of May 9, 2007:
http://www.ustreas.gov/offices/domestic-finance/financial-institution/terrorism-insurance/

U.S. General Accounting Office, *Terrorism Insurance: Rising Uninsured Exposure to Attacks Heightens Potential Economic Vulnerabilities,* Washington, D.C.: U.S. General Accounting Office, GAO-02-472T, February 27, 2002. As of September 6, 2007:
http://purl.access.gpo.gov/GPO/LPS38448

U.S. Government Accountability Office, and U.S. House of Representatives, Committee on Financial Services, *Terrorism Insurance: Measuring and Predicting Losses from Unconventional Weapons Is Difficult, but Some Industry Exposure Exists: Report to the Chairman, Committee on Financial Services, House of Representatives,* Washington, D.C.: U.S. Government Accountability Office, GAO-06-1081, September 2006. As of September 6, 2007:
http://purl.access.gpo.gov/GPO/LPS76534

U.S. House of Representatives, "A Bill to Extend the Terrorism Insurance Program of the Department of the Treasury, and for Other Purposes," H.R. 2761, 110th Congress, 1st Session, June 18, 2007.

U.S. Small Business Administration, "Fact Sheet About U.S. Small Business Administration Disaster Loans," Disaster Area 1 Office, Niagara Falls, N.Y., 2002.